HELMINTH ZOONOSES

CURRENT TOPICS IN VETERINARY MEDICINE AND ANIMAL SCIENCE

Control of Reproduction in the Cow, edited by J.M. Sreenan

Patterns of Growth and Development in Cattle, edited by H. de Boer and J. Martin

Respiratory Diseases in Cattle, edited by W.B. Martin

Calving Problems and Early Viability of the Calf, edited by B. Hoffmann, I.L. Mason and J. Schmidt

The Future of Beef Production in the European Community, edited by J.C. Bowman and P. Susmel

Diseases of Cattle in the Tropics: Economic and Zoonotic Relevance, edited by M. Ristic and I. McIntyre

Control of Reproductive Functions in Domestic Animals, edited by W. Jöchle and D.R. Lamond

The Laying Hen and its Environment, edited by R. Moss

Epidemiology and Control of Nematodiasis in Cattle, edited by P. Nansen, R.J. Jørgensen and E.J.L. Soulsby

The Problem of Dark-Cutting in Beef, edited by D.E. Hood and P.V. Tarrant

The Welfare of Pigs, edited by W. Sybesma

The Mucosal Immune System, edited by F.J. Bourne

Laboratory Diagnosis in Neonatal Calf and Pig Diarrhoea, edited by P.W. de Leeuw and P.A.M. Guinée

Advances in the Control of Theileriosis, edited by A.D. Irvin, M.P. Cunningham and A.S. Young

Fourth International Symposium on Bovine Leukosis, edited by O.C. Straub

Muscle Hypertrophy of Genetic Origin and its Use to Improve Beef Production, edited by J.W.B. King and F. Ménissier

Aujeszky's Disease, edited by G. Wittman and S.A. Hall

Transport of Animals Intended for Breeding, Production and Slaughter, edited by R. Moss

Welfare and Husbandry of Calves, edited by J.P. Signoret

Factors Influencing Fertility in the Postpartum Cow, edited by H. Karg and E. Schallenberger

Beef Production from Different Dairy Breeds and Dairy Beef Crosses, edited by G.J. More O'Ferrall

The Elisa: Enzyme-Linked Immunosorbent Assay in Veterinary Research and Diagnosis, edited by R.C. Wardley and J.R. Crowther

Indicators Relevant to Farm Animal Welfare, edited by D. Smidt

Farm Animal Housing and Welfare, edited by S.H. Baxter, M.R. Baxter and J.A.D. MacCormack

Stunning of Animals for Slaughter, edited by G. Eikelenboom

Manipulation of Growth in Farm Animals, edited by J.F. Roche and D. O'Callaghan

Latent Herpes Virus Infections in Veterinary Medicine, edited by G. Wittmann, R.M. Gaskell and H.-J. Rziha

Grassland Beef Production, edited by W. Holmes

Recent Advances in Virus Diagnosis, edited by M.S. McNulty and J.B. McFerran

The Male in Farm Animal Reproduction, edited by M. Courot

Endocrine Causes of Seasonal and Lactational Anestrus in Farm Animals, edited by F. Ellendorff and F. Elsaesser

Brucella Melitensis, edited by J.M. Verger and M. Plommet

Diagnosis of Mycotoxicoses, edited by J.L. Richard and J.R. Thurston

Embryonic Mortality in Farm Animals, edited by J.M. Sreenan and M.G. Diskin

Social Space for Domestic Animals, edited by R. Zayan

The Present State of Leptospirosis Diagnosis and Control, edited by W.A. Ellis and T.W.A. Little

Acute virus infections of poultry, edited by J.B. McFerran and M.S. McNulty

Evaluation and Control of Meat Quality in Pigs, edited by P.V. Tarrant, G. Eikelenboom and G. Monin

Follicular Growth and Ovulation Rate in Farm Animals, edited by J.F. Roche and D. O'Callaghan

Cattle Housing Systems, Lameness and Behaviour, edited by H.K. Wierenga and D.J. Peterse

Physiological and Pharmacological Aspects of the Reticulo-Rumen, edited by L.A.A. Ooms, A.D. Degryse and A.S.J.P.A.M. van Miert

Biology of Stress in Farm Animals: An Integrative Approach, edited by P.R. Wiepkema and P.W.M. van Adrichem

Helminth Zoonoses, edited by S. Geerts, V. Kumar and J. Brandt

HELMINTH ZOONOSES

Edited by

S. Geerts, V. Kumar and J. Brandt
Department of Veterinary Medicine, Institute of Tropical Medicine
'Prins Leopold', Antwerp, Belgium

1987 **MARTINUS NIJHOFF PUBLISHERS**
a member of the KLUWER ACADEMIC PUBLISHERS GROUP
DORDRECHT / BOSTON / LANCASTER

Distributors

for the United States and Canada: Kluwer Academic Publishers, P.O. Box 358, Accord Station, Hingham, MA 02018-0358, USA
for the UK and Ireland: Kluwer Academic Publishers, MTP Press Limited, Falcon House, Queen Square, Lancaster LA1 1RN, UK
for all other countries: Kluwer Academic Publishers Group, Distribution Center, P.O. Box 322, 3300 AH Dordrecht, The Netherlands

Library of Congress Cataloging in Publication Data

Helminth zoonoses.

    (Current topics in veterinary medicine and animal
science)
    Proceedings of the International Colloquium on Helminth
Zoonoses held at the Institute of Tropical Medicine,
Antwerp, 11-12 December 1986.
    Includes index.
    1. Helminthiasis--Congresses. 2. Zoonoses--Congresses.
I. Geerts, S. (Stanny) II. Kumar, V. (Vinai) III. Brandt,
J. (Jef) IV. International Colloquium on Helminth
Zoonoses (1986 : Antwerp, Belgium) V. Series.
RC119.H45 1987   616.9'62   87-12315

ISBN-13: 978-94-010-8001-9   e-ISBN-13: 978-94-009-3341-5
DOI: 10.1007/978-94-009-3341-5

Copyright

PREFACE

In spite of the availability of modern broad-spectrum anthelmintic
drugs, the prevention and control of helminth zoonoses remain a challenge
to human and veterinary parasitologists and to physicians and veterinarians working on the field. Although the life cycles of most helminths of
zoonotic importance are well known, there are still major gaps in our
knowledge especially in the fields of epidemiology, diagnosis and treatment

The International Colloquium on Helminth Zoonoses held at the
Institute of Tropical Medicine, Antwerp, 11-12 December 1986, laid emphasis
on more recent advances made in the control and epidemiology of these
zoonotic diseases. The disease complexes echinococcosis/hydatidosis,
taeniasis/cysticercosis and the larva migrans-syndrome were dealth with in
considerable detail.

In the first chapter the phenomenon of strain variation in
Echinococcus spp. is examined in the light of newer findings. The progress
made in recent years towards a more specific diagnosis and drug targeting
in hydatidosis is reported. In the second chapter recent advances in
immunisation and treatment of cysticercosis are dealt with. The possibility of the existence of strain differences in Taenia saginata is also discussed. The third chapter is devoted to trematode zoonoses with particular
reference to the situation in South-east Asia, Senegal (schistosomiasis)
and Liberia (paragonimiasis). In the last chapter the larva migrans-
syndrome is treated in detail with special attention to its etiology and
and diagnosis. Reports on lesser known nematode zoonoses like mammomonogamosis and oesophagostomiasis are included.

The editors hope that the present proceedings, emcompassing not only
the recent advances in the field of zoonotic helminths and helminthiasis
but also highlighting some of the serious gaps in our present day knowledge, shall further stimulate the readers in their endeavour for a better
understanding and more efficient control of these diseases which still
constitute a menace to the public health and livestock industry.

Special thanks are due to the staff of the veterinary department, the
secretariate and the workshops of the Institute of Tropical Medicine,
Antwerp for their help in the organisation of the Colloquium and the
preparation of these proceedings.

Stanny Geerts
Vinai Kumar
Jef Brandt

CONTENTS

Preface   V

List of contributors   XI

Section 1: Echinococcosis - Hydatidosis

Changing concepts in the microecology, macroecology and epidemiology of
hydatid disease
    J.D. Smyth   1

The fertility of hydatid cysts in food animals in Greece
    C.H. Himonas, S. Frydas and K. Antoniadou-Sotiriadou   12

A failure to infect dogs with Echinococcus granulosus protoscoleces of
human origin
    T. Wikerhauser, J. Brglez, M. Stulhofer and B. Ivanisevic   22

Experimental infection of sheep and monkeys with the camel strain of
Echinococcus granulosus
    G. Macchioni, M. Arispici, P. Lanfranchi and F. Testi   24

Characterization of the hydatid disease organism, Echinococcus granulosus
from Kenya using cloned DNA markers
    D.P. McManus, A.J.G. Simpson and A.K. Rishi   29

Kinetics of molecular transfer across the tegument of protoscoleces and
hydatid cysts of Echinococcus granulosus and the relevance of these studies
to drug targeting
    S.A. Jeffs, H. Hurd, J.T. Allen and C. Arme   37

Diagnosis of ovine hydatidosis by immunoelectrophoresis
    F. Martinez-Gomez, S. Hernandez-Rodriguez and F. Serrano-Aguilera   44

Characterisation of Echinococcus granulosus proteins and antigens from
hydatid cyst fluid
    J.C. Shepherd and D.P. McManus   50

Existence of an urban cycle of Echinococcus granulosus in Central Tunisia
    A. Bchir, A. Jaiem, M. Jemmali, J.J. Rousset, L. Gaudebout and
    B. Larouze   57

Echinococcosis eradication in Cyprus
    K. Polydorou   60

VIII

Section 2: Taeniasis - Cysticercosis

Taeniasis: the tantalizing target
M.M.H. Sewell   68

Observations on possible strain differences in Taenia saginata
G. Wouters, J. Brandt and S. Geerts   76

Immunoprophylaxis of Taenia saginata cysticercosis
L. Harrison and G.W.P. Joshua   81

An important focus of porcine and human cysticercosis in West Cameroon
A. Zoli Paanah, S. Geerts and T. Vervoort   85

Cysticercus fasciolaris in mice: a laboratory model for selecting new drugs
on cysticercosis
O. Vanparijs, L. Hermans and H. Lauwers   92

Large-scale use of chemotherapy of taeniasis as a control measure for
T. solium infections
Z.S. Pawlowski   100

Section 3: Trematode Zoonoses

Zoonotic trematodiasis in South-east and Far-east Asian countries
V. Kumar   106

Observations on human and animal schistosomiasis in Senegal
D. Rollinson, J. Vercruysse, V.R. Southgate, P.J. Moore, G.G. Ross,
T.K. Walker and R.J. Knowles   119

Occurrence of human lung fluke infection in an endemic area in Liberia
S. Sachs   132

Section 4: Larva migrans and other nematode zoonoses

Larva migrans in perspective
E.J.L. Soulsby   137

Immunological studies on Ascaris suum infections in mice
P. van Than and F. van Knapen   149

Toxocara vitulorum: a possible agent of larva migrans in humans?
K. van Gorp, M. Mangelschots and J. Brandt   159

Antigenic and biochemical analysis of the E.S. molecules of Toxocara canis
infective larvae
B.D. Robertson, S. Rathaur and R.M. Maizels   167

Ocular toxocariasis: role of IgE in the pathogenesis of the syndrome and
diagnostic implications
C. Genchi, P. Falagiani, G. Riva and C. Sioli   175

Serological arguments for multiple etiology of visceral larva migrans
  J.C. Petithory, F. Derouin, M. Rousseau, M. Luffau and M. Quedoc   183

Experimental Trichinella spiralis infection in two horses
  F. van Knapen, J.H. Franchimont, W.M.L. Hendrickx and M. Eysker   192

Intestinal mast cells: possible regulation and their function in the gut of
Trichinella spiralis infected small rodents
  J. Buys, H.K. Parmentier, H. van Loveren and J. Ruitenberg   202

Variable levels of host immunoglobulin on microfilariae of Brugia pahangi
isolated from the blood of cats
  U.N. Premaratne, R.M.E. Parkhouse and D.A. Denham   215

Mammomonogamosis
  J. Euzeby, J. Gevrey, M. Graber and A. Mejia-Garcia   225

Frequency of symptomatic human oesophagostomiasis (helminthoma) in
Northern Togo
  P. Gigase, S. Baeta, V. Kumar and J. Brandt   228

Index of Subjects   237

LIST OF CONTRIBUTORS

ALLEN JT, Parasitology Research Laboratory, University of Keele,
Keele, Staffordshire ST5 5BG, U.K.

ANTONIADOU-SOTIRIADOU K, Department of Infectious and Parasitic Diseases,
Avian Medicine and Pathology, Faculty of Veterinary Medicine,
Aristotelian University, Thessaloniki 540 06, Greece.

ARISPICI M, Dipartimento di Patologia Animale, Profilassi e Igiene degli
Alimenti, Università di Pisa, V.le delle Piagge, 2, 56100 Pisa, Italy.

ARME C, Parasitology Research Laboratory, University of Keele, Keele,
Staffordshire ST5 5BG, U.K.

BAETA S, Formerly Chief Medical Officer, District Hospital of Dapaong, Togo.

BCHIR A, Facultes de medecine de Sousse et de Monastir, Tunesie.

BRANDT J. Institute of Tropical Medicine, Nationalestraat 155,
2000 Antwerp, Belgium.

BRGLEZ J, Veterinary Division of the University of Ljubljana, Gerbičeva 60,
61000 Ljubljana, Yugoslavia.

BUIJS J, National Institute of Public Health and Environmental Hygiene,
P.O.Box 1, 3720 BA Bilthoven, The Netherlands.

DENHAM DA, London School of Hygiene and Tropical Medicine, Keppelstreet,
London WC1E 7HT, U.K.

DEROUIN F, Service de Biologie Médicale Emile Brumpt, Centre Hospitalier,
95500 Gonesse, France.

EUZEBY J, Ecole Nationale Vétérinaire de Lyon, B.P.31, 69752 Charbonnières,
Cedex, France.

EYSKER M, National Institute of Public Health and Environmental Hygiene,
P.O.Box 1, 3720 BA Bilthoven, The Netherlands.

FALAGIANI P, Reparto ricerche, Laboratorio Lofarma, viale Cassala 40,
20143 Mila, Italy.

FRANCHIMONT JH, National Institute of Public Health and Environmental
Hygiene, P.O.Box 1, 3720 BA Bilthoven, The Netherlands.

FRYDAS S, Department of Infectious and Parasitic Diseases, Avian Medicine
and Pathology, Faculty of Veterinary Medicine, Aristotelian
University, Thessaloniki 540 06, Greece.

GAUDEBOUT C, INSERM U13/IMEA, Hopital Claude Bernard, 75944 Paris,
Cedex 19, France.

GEERTS S, Institute of Tropical Medicine, Veterinary Department,
Nationalestraat 155, 2000 Antwerp, Belgium.

GENCHI C, Istituto di Patologia Generale Veterinaria, via Celoria 10, 20133 Milan, Italy.

GEVREY J, Ecole Nationale Vétérinaire de Lyon, B.P.31, 69752 Charbonnières, Cedex, France.

GIGASE P, Institute of Tropical Medicine, B-2000 Antwerp, Belgium.

GRABER M, Ecole Nationale Vétérinaire de Lyon, B.P.31, 69752 Charbonnières, Cedex, France.

HARRISON LJS, University of Edinburgh, Centre for Tropical Veterinary Medicine, Easter Bush, Roslin, Midlothian, Scotland, EH25 9RG.

HENDRIKX WML, Veterinary Faculty, University of Utrecht, Yalelaan 7, 3584 CL Utrecht, The Netherlands.

HERMANS L, Department of Parasitology, Janssen Pharmaceutica, Turnhoutseweg 30, 2340 Beerse, Belgium.

HERNANDEZ-RODRIGUEZ S, Department of Parasitology and Parasitic diseases. Veterinary Faculty. Av.Medina Ahzahara, 14005 Córdoba, Spain.

HIMONAS C, Department of Infectious and Parasitic Diseases, Avian Medicine and Pathology, Faculty of Veterinary Medicine, Aristotelian University, Thessaloniki 540 06, Greece.

HURD H, Parasitology Research Laboratory, University of Keele, Keele, Staffordshire ST5 5BG, U.K.

IVANISEVIC B, Department of Surgery of the Medical Faculty, Rebro, 41000 Zagreb, Yugoslavia.

JAIEM A, Facultes de medecine de Sousse et de Monastir, Tunisie.

JEFFS SA, Parasitology Research Laboratory, University of Keele, Keele, Staffordshire ST5 5BG, U.K.

JEMMALI M, Facultes de medecine de Sousse et de Monastir, Tunisie.

JOSHUA GWP, University of Edinburgh, Centre for Tropical Veterinary Medicine, Easter Bush, Roslin, Midlothian, Scotland, EH25 9RG.

KNOWLES RJ, Department of Zoology, British Museum (Natural History), London SW7 5BD, U.K.

KUMAR V, Institute of Tropical Medicine, Veterinary Department, Nationalestraat 155, B-2000 Antwerp, Belgium.

LANFRANCHI P, Dipartimento di Patologia Animale, Profilassi e Igiene degli Alimenti, Università di Pisa, V.le delle Piagge 2, 56100 Pisa, Italy.

LAROUZE B, INSERM U13/IMEA, Hopital Claude Bernard, 75944 Paris, Cedex 19, France.

LAUWERS H, Department of Parasitology, Janssen Pharmaceutica, Turnhoutseweg 30, 2340 Beerse,

LUFFAU M, I.N.A., 78850 Grignon, France .

MACCHIONI G, Dipartimento di Patologia Animale, Profilassi e Igiene degli Alimenti, Università di Pisa, V.le delle Piagge, 2, 56100 Pisa, Italy.

McMANUS DP, Department of Pure and Applied Biology, Imperial College
of Science and Technology, Prince Consort Road, London SW7 2BB, U.K.

MAIZELS RM, Department of Pure and Applied Biology, Imperial College,
London SW7, U.K.

MANGELSCHOTS M, Institute of Tropical Medicine, Nationalestraat 155,
B-2000 Antwerp, Belgium.

MARTINEZ-GOMEZ F, Department of Parasitology and Parasitic diseases.
Veterinary Faculty, Av.Medina Ahzahara, 14005 Córdoba, Spain.

MEJIA-GARCIA A, Ecole National Vétérinaire de Lyon, B.P. 31,
69752 Carbonnières, Cedex, France.

MOORE PJ, Department of Zoology, British Museum (National History),
London SW7 5BD, U.K.

PARKHOUSE RMF, National Institute for Medical Research, The Ridgeway,
Mill Hill, London NW7 1AA, U.K.

PARMENTIER HK, Department of Immunology, Faculty of Veterinary Medicine,
University of Utrecht, Yalelaan 1, Utrecht, The Netherlands.

PAWLOWSKI ZS, Parasitic Diseases Programme, World Health Organization,
1211 Geneva 27, Switzerland.

PETITHORY JC, Service de Biologie Médicale Emile Brumpt, Centre
Hospitalier, 95500 Gonesse, France.

POLYDOROU K, Director, Department of Veterinary Services, Nicosia, Cyprus.

PREMARATNE UN, London School of Hygiene and Tropical Medicine,
Keppelstreet, London WC1E 7HT, U.K.

QUEDOC M, Service de Biologie Médicale Emile Brumpt, Centre Hospitalier,
95500 Gonesse, France.

RATHAUR S, Department of Pure & Applied Biology, Imperial College,
London SW7, U.K.

RISHI AK, Department of Pure and Applied Biology, Imperial College of
Science and Technology, Prince Consort Road, London SW7 2BB, U.K.

RIVA G, Reparto ricerche, Laboratoirio Lofarma, viale Cassala 40,
20143 Mila, Italy.

ROBERTSON BD, Department of Pure & Applied Biology, Imperial College,
London SW7, U.K.

ROLLINSON D, Department of Zoology, British Museum (Natural History),
London SW7 5BD, U.K.

ROSS GC, Department of Zoology, British Museum (Natural History),
London SW7 5BD, U.K.

ROUSSEAU M, Service de Biologie Médicale Emile Brumpt, Centre Hospitalier,
95500 Gonesse, France

ROUSSET JJ, Departement de Parasitologie, UER Medecine, Bobigny, France

RUITENBERG EJ, National Institute of Public Health and Environmental
Hygiene, P.O.Box 1, 3720 BA Bilthoven, The Netherlands.

SACHS R, Department of Veterinary Medicine, Bernhard-Nocht-Institute for Maritime and Tropical Diseases, D-2000 Hamburg 4, F.R. Germany.

SERRANO-AGUILERA F, Department of Parasitology and Parasitic diseases. Veterinary Faculty. Av.Medina Ahzahara, 14005 Córdoba, Spain.

SEWELL MMH, University of Edinburgh, Centre for Tropical Veterinary Medicine, Easter Bush, Roslin, Midlothian Scotland, EH25 9RG.

SHEPHERD JC, Department of Pure and Applied Biology, Imperial College of Science and Technology, Prince Consort Rd., London SW7 2BB, U.K.

SIMPSON AJG, Division of Parasitology, National Institute for Medical Research, The Ridgeway, Mill Hill, London NW7 1AA, U.K.

SIOLI C, Instituto di Patologia Generale Veterinaria, via Celoria 10, 20133 Milan, Italy.

SMYTH JD, Department of Medical Helminthology, London School of Hygiene and Tropical Medicine, Keppell st. (Gower St.), London WC1E 7HT,U.K.

SOULSBY ESL, University of Cambridge, Department of Clinical Veterinary Medicine,Madingley Road, Cambridge CB3 OES, U.K.

SOUTHGATE VR, Department of Zoology, British Museum (Natural History), London SW7 5BD, U.K.

STULHOFER M, Department of Surgery, University Hospital, Zajčeva 19, 41000 Zagreb, Yugoslavia .

TESTI F, Dipartimento di Patologia Animale, Profilassi e Igiene degli Alimenti, Università di Pisa, V.le delle Piagge, 2, 56100 Pisa, Italy.

VAN GORP K, Institute of Tropical Medicine, Nationalestraat 155, B-2000 Antwerp, Belgium.

VAN KNAPEN F, National Institute of Public Health and Environmental Hygiene, P.O.Box 1, 3720 BA Bilthoven, The Netherlands.

VAN LOVEREN H, National Institute of Public Health and Environmental Hygiene, P.O.Box 1, 3720 BA Bilthoven, The Netherlands.

VANPARIJS O, Department of Parasitology, Janssen Pharmaceutica, Turnhoutse-weg 30, 2340 Beerse, Belgium.

VAN THAN P, National Institute of Public Health and Environmental Hygiene, P.O.Box 1, 3720 BA Bilthoven, The Netherlands.

VERCRUYSSE J, Faculteit Diergeneeskunde, Laboratorium voor Parasitologie en Parasitaire Ziekten, Casinoplein 24, B-9000 Gent, Belgium.

VERVOORT T, Institute of Tropical Medicine, Nationalestraat 155, B-2000 Antwerp, Belgium.

WALKER TK, Department of Zoology, British Museum (Natural History), London SW7 5BD, U.K.

WIKERHAUSER T, Veterinary Faculty, Heinzelova 55, 41000 Zagreb,Yugoslavia.

WOUTERS G, Institute of Tropical Medicine, Veterinary Department, Nationalestraat 155, B-2000 Antwerp, Belgium.

ZOLI A, Centre Universitaire de Dschang, B.P.110, Cameroun.

CHANGING CONCEPTS IN THE MICROECOLOGY, MACROECOLOGY AND EPIDEMIOLOGY OF
HYDATID DISEASE.

J.D. SMYTH

ABSTRACT
The ecology of hydatid disease can be considered to involve the interaction between two biological systems: those involving the intimate relationship of Echinococcus granulosus or E.multilocularis with the microenvironment of their definitive and intermediate hosts, which represents the
microecology of these organisms and the inter-relationship of these definitive and intermediate hosts, both between themselves and with their
external environment, which represents the macroecology. Understanding the
nature of the host-parasite relationship, and hence the epidemiology of
the disease, involves a detailed knowledge of both these interacting systems. Recent advances in selected areas of the developmental biology of
Echinococcus spp. in vivo and in vitro are reviewed, drawing attention to
areas where gaps in our knowledge exist and where further work is particularly needed. Studies on the microecology have led to the now well-recog-
nised concept that E.granulosus and (probably E.multilocularis) exist as
a number of different strains, and the mechanism whereby these may originate and their significance is examined. The impact of these studies on
our understanding of the overall ecology and epidemiology of hydatid
disease and its control is discussed.

1. INTRODUCTION
This paper attempts to review how knowledge of the biology of the cestodes,
Echinococcus granulosus and E.multilocularis, has changed within recent
years and the impact this has had on our understanding of the epidemiology
of hydatid disease (hydatidosis, echinococcosis).
The biology of the organisms can be best examined by reminding ourselves
that, like many internal parasites, their life cycles involve two interacting ecosystems with markedly different environments, (a) the external
(free-living) environment in which the potential definitive and intermediate hosts live - the study of which can be termed the macroecology, and
(b) the internal or microenvironment provided by the organ systems of
these hosts - the study of which can be termed the microecology. In order
to obtain an overall picture of the host-parasite relationship - and hence
of the epidemiology of the resulting disease - it is clearly necessary to
have an understanding of both these interacting systems.
Because so much has been written about the biology of hydatid disease
within recent years (30, 31) the scope of this review has been limited to
(a) research carried out within the last decade on the basic biology of
E.granulosus and E.multilocularis and (b) the influence of this new knowledge on our understanding of the epidemiology of hydatid disease . and its
control.

2. MICROECOLOGY OF ECHINOCOCCUS SPP.

The physiological features of the stages of the life cycle within the definitive and intermediate host, which essentially represent the microecological features of the life cycle, are shown in Fig.1. Recent research on each stage is considered below.

2.1. Adult worms

Relatively little progress has been made in our knowledge of the association of the adult worm with the carnivore gut. It has been known for a long time that the scolex of both E.granulosus and E.multilocularis make intimate contact with the crypts of Lieberkühn in the intestine the cells of which become flattened and eroded at the site of contact. It is also known that scoleces of both species secrete a labile substance (possibly a lipoprotein) from a group of gland cells in the rostellum (the rostellar gland) (20, 26, 32, 33). This secretion and/or other E/S antigens are probably responsible for the appearance of antibody in the sera, which appears some 14 days after infection (3). This serological test is likely to prove of value in epidemiological surveys of dog populations and could be especially useful in testing the efficacy of control programmes.

Within recent years, more information has been obtained regarding the prepatent period of E.granulosus and E.multilocularis, a knowledge of which is essential in the organization of an efficient control programme. For many years, it has been accepted that the prepatent period of E.granulosus in the dog was aproximately 38-40 days (19). This led to the supposition that it was safe to maintain and handle dogs infected with E.granulosus for up to 35 days p.i. without risk of infection. This assumption has been shown to be no longer tenable by the discovery that in Switzerland the cattle/dog isolate of this species could become ovigerous as early as 35 days (34). Other variations on the prepatent period are now beginning to emerge from other countries. For example, in Australia, isolates of E.granulosus from different areas have been reported to have different prepatent periods, the organisms of Tasmanian origin producing eggs approximately 7 days earlier than those of Eastern or Western Australian origins (6).

Again, an unexpected finding in Russia has been that the maturation of E.granulosus in dogs was longer (39-54 days) in summer than in winter (35-49 days) (35). In another Russian study, it was found that cysts from wild boars fed to dogs gave rise to ovigerous worms in 45-47 days (18). It is possible that the external environmental temperature may have an effect on the maturation time in dogs and this could be important, perhaps, in the tropics, although this question does not appear to have been examined.

A further remarkable observation (at present unconfirmed) has been reported from India, where it was found that hydatid cysts from different organs (e.g.liver, lungs) from the same host gave rise to adults in dogs with different prepatent periods (Nizami, personal observation.) This extraordinary result suggests that the organ origin of cyst as well the host origin may have to be taken into account when the prepatent period is being considered. It could be, that in different sites the nutrition available differs sufficiently to affect the development of the adult worm and that protoscoleces with "better" food reserves may be able to develop more rapidly than those with "poor" reserves, which may need a longer time to acquire the nutritional factors necessary for differentiation and maturation.

3

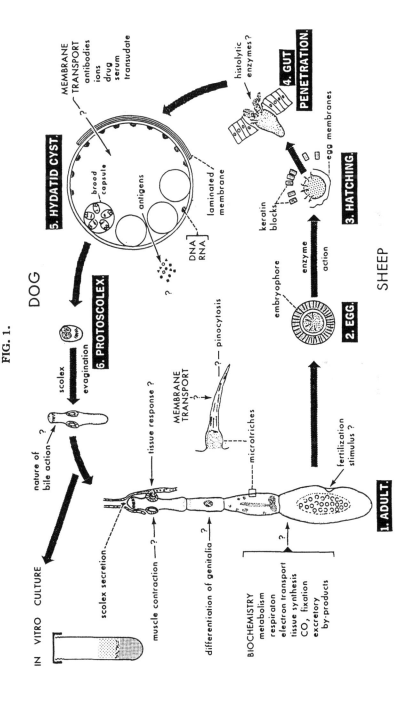

FIG. 1.

Fig. 1. Microecological aspects of the life cycle of Echinococcus granulosus. The question marks indicate areas of particular importance where further research is needed.

Much less information is available on the maturation of E.multilocularis, presumably on account of the difficulties and dangers of maintaining this species to patency in dogs. It has long been assumed that the prepatent period for this species has been about 30 days (19) but more recent work, based on in vivo followed by in vitro studies suggests a range of 28-35 days (33).

The chief definitive host of E.multilocularis is recognised to be the fox, with the dog playing a lesser role. Very limited studies have been carried out on the role of the cat as a definitive host, a situation which could be of special importance in urban situations. A recent study (5) with the Alaskan isolate of E.multilocularis found that in cats there was a marked rejection of established strobila after 10 days and the growth was very retarded compared with that in dogs, as evidenced by the slow increase in lenght as well as the late formation of the posterior proglottid.

2.2. Egg

Our knowledge of the biology of the eggs has only increased a little within recent years. This, again, undoubtedly reflects the difficulties and hazards of obtaining viable eggs of Echinococcus and experimenting with them. One very important advance made by Dr.P.Craig and his colleagues (1) has been the development of an immunological method for the specific iden-tification of eggs of Echinococcus. It has long been recognised that a fundamental problem in the control of hydatid disease has been the inabi-lity to distinguish the eggs of Echinococcus from those of other common Taeniidae of dogs, such as Taenia ovis. This method involves fairly complex immunological procedures and it remains to be seen if the method can be sufficiently simplified to be of value for routine use in diagnostic labo-ratories or field situations.

A further useful contribution to our knowledge of the structure of the egg has been the observation of Swiderski (29) that in E.granulosus the so-called "penetration" glands consist of 3 different types of gland cells. Considering that the E/S antigens of oncospheres have been widely used as potential vaccines, this is an valuable observation and it is clearly im-portant that the secretions of these glands be isolated and their compo-sition determined.

2.3. Hydatid cyst

The structure and general biology of the hydatid cyst has received a dis-proportionate amount of attention relative to the rest of the life cycle and this topic will only be briefly considered here. Much of the emphasis, within the last decade, has centred on the effect of various drugs on the viability of the cysts and their use as a substitute for, or an adjunct to, surgery. This topic has been extensively reviewed (14, 17) and of the two drugs most extensively used, mebendazole and albendazole, the latter is emerging as the most promising as regard to its clinical efficacy, non-toxicity and general lack of side effects.

The chemotherapeutic studies have raised two important problems on the basic biology of the hydatid cyst. (a) Why do these drugs, while frequently killing the protoscoleces, usually fail to kill the germinal membrane and (b) apart from the work of Reisin (15) and his colleagues (see below), relatively little is known regarding the penetration of substances (such as drugs, nutrients or antibodies) into the hydatid cyst - especially in men.

Although the structure and ultrastructure of the cyst has been extensively studied,surprisingly little is known regarding the structure and physiology of the germinal membrane. Knowledge of this may prove to be an important consideration in postoperative procedures to determine whether surgically-removed cyst tissue is alive or not. If intact protoscoleces are recovered some indication can be obtained by testing with (a) the eosin exclusion test (whereby living protoscoleces do not permit entry of the stain, whereas dead ones stain); this test is recognised as being very subjective and unreliable; (b) flame cell activity - reliable; (c) response to bile or bile salts - muscular movement and/or evagination occurs - reliable; (d) pepsin treatment, which may be followed by in vitro culture (24) - live protoscoles survive pepsin digestion - reliable; (e) rodent injection-reliable.
Of all the above tests, rodent injection is the only absolutely reliable test, as the appearance of secondary cysts is unequivocal proof of the presence of living cysts material and should be used as a control if other tests prove to be negative.
One of the major difficulties recognised in hydatid research is the lack of suitable laboratory material. The findings of Eckert & Ramp (2) that cryopreservation of cystic E. multilocularis is possible represents a valuable step forward. They developed a technique involving 10% glycerol as a cryoprotectant and a three-step freezing schedule. This technique should be of great value in furthering research in multilocular hydatid disease. As mentioned earlier, the physico-chemistry of hydatid cysts has been very little studied and it is therefore important to note that in Argentina, Dr. Reisin and his colleagues have carried out fundamental studies on the transport of drugs, carbohydrates and electrolytes into hydatid tissues (16, 15).

3. MACROECOLOGY
A study of macroecology of the disease involves knowledge of the ecology of the potential definitive and intermediate hosts. This is a vast topic and one which can only be touched on briefly here.
3.1. E.granulosus
It is well known that numerous species of carnivores can act as definitive hosts for this species (19) and some 70 species of ungulates can act as intermediate hosts.
It is also now widely recognised from in vitro, in vivo, biochemical and morphological studies (6, 7, 9, 12, 25) that isolates from different hosts or the same host from different areas may have different characteristics. This has led to the concept that E.granulosus exists in a number of different strains, not all of which may be infective to man or cross-infective to each other. On account of the experimental difficulties, very few cross-infection experiments have been carried out. Perhaps, the most studied isolates has been those from horse and sheep which have been shown to have marked morphological, biochemical, nutritional and developmental differences (9, 12). Although some 60% of horses in the United Kingdom have been reported to be infected, circumstantial evidence suggests that the horse isolate may not be infective to man (13, 21).
Epidemiological evidence suggests that E.granulosus of sheep, camel, pig are all infective to man, but even within the same host in the same country, different strains have been reported (7). Possible differences between

FIG. 2.

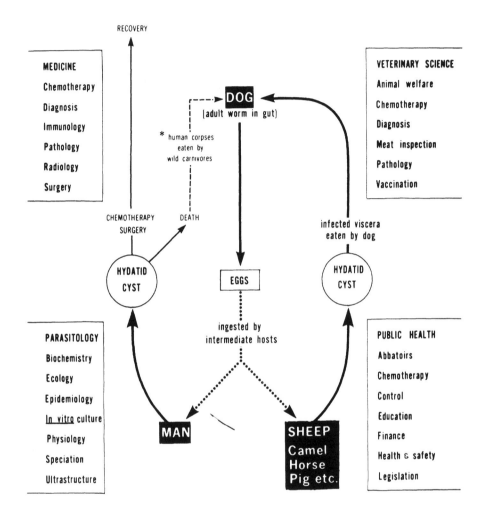

Fig. 2. Macroecology of hydatid disease: the macroecological aspects of the life cycle of Echinococcus granulosus and the various disciplines involved in the study of the disease and its control. *Refers to the exceptional situation in Kenya where, on death, human corpses are left in the bush to be eaten by wild carnivores.

strains which may be important in control programmes are factors such as
the prepatent period, reactions to drugs, immunological responses and
metabolic differences.
Effective control programmes involve application of many disciplines and
the interactions of many of these are shown in Fig.2.
3.2. E.multilocularis.
The question of whether or not strains of E.multilocularis also occur is
not clear and further work along the lines carried out on E.granulosus -
especially on the biochemistry and developmental biology of various isolates - is clearly needed.

4. NICHE THEORY AND THE "STRAIN"COMPLEX

Consideration of "Niche" Theory as applied to parasites may provide some
theoretical background to the appearance of strains of E.granulosus and
E.multilocularis (23). All living organisms can be said to occupy a biological "niche", but few terms have been so misunderstood and yet it is a
concept of considerable significance to parasitologists. It essentially
refers to a "space" in the biotic environment in which life is possible.
Some workers use the term in a restrictive sense - confusing it with the
term "habitat" - which is really only the environmental component of the
niche. Others speak of "feeding niches", i.e. one species using one type of
food and another a different type, in the same habitat. Other terms, such
as "behavioural niches", "metabolic niches" are also used.
These considerations have given rise to the concept of a fundamental niche,
which has been defined as "that unique combination of environmental factors
biotic and abiotic, which are capable of supporting life". Since there are
an unlimited number of both biotic and abiotic factors to be considered we
arrive at the (abstract) concept of a fundamental or ecological niche as an
n-dimensional hypervolume.
We can ask the question, what happens when two species (or populations)
attempt to occupy the same niche. This problem was considered by the
Russian biologist Gause who postulated what is now known (4) as Gause's
Hypothesis (Law, Principle) which states: "No two forms can occupy the
same ecological niche for an indefinite period, eventually one form will
replace the other".
This hypothesis has been restated by Hardin (4) as the "Competitive Exclusion Principle" which simply states: "Complete competitors cannot co(exist".
The implications of this hypothesis are shown in Fig.3.
If we examine the niches occupied (for example) by the sheep and horse
strains respectively (Fig.4) we can see these overlap in their sharing of
the environment provided by the dog gut. According to the above hypothesis,
if the overlaping of the niches is substantial, one strain could be expected to replace the other. However, if these strains have evolved different
metabolic pathways, their niches would become separated and competition for
the same nutrients in the dog gut would be reduced, thus enhancing the
chance of both surviving.
This, in fact, appears to be exactly what has happened in nature, the
horse and sheep isolates have evolved differen metabolisms and different
nutritional requirements (9, 10, 11) thus separating their niches sufficiently to allow both strains to survive.

8

FIG. 3.

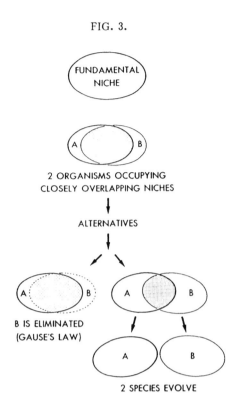

Fig.3. Gause's Hypothesis. If two species attempt to occupy the same "niche" at the same time, either one will replace the other, or the species will evolve in such a way that their niches no longer overlap to a significant extent.

5. GENETICAL BASIS FOR STRAINS DIFFERENCES

Theoretically, the evolution of new strains in Echinococcus spp. probably take place more readily than in most other species of cestodes, due to the fact that (a) the adult is a hermaphrodite and undergoes self-fertilization (8, 28) and hence a single recessive mutation could appear in both sperm and ova at the same time, thus increasing the likelihood of a double recessive mutant appearing and (b) the single egg, by polyembryony, gives rise to large numbers of embryos (protoscoleces) in a hydatid cyst. Hence, if a mutant is successful in penetrating and surviving in an intermediate host, it can produce many thousands of genetically identical individuals (i.e. a clone) and a new strain has made its appearance (21, 27).

9

FIG.4.

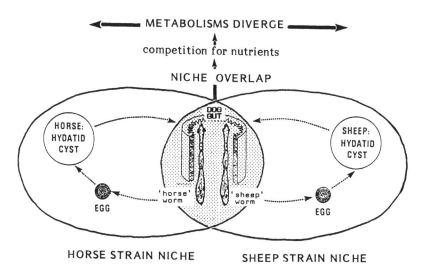

HORSE STRAIN NICHE   SHEEP STRAIN NICHE

Fig.4. Gause's Hypothesis applied to the sheep and horse "strains"of
Echinococcus granulosus. The niches of these two strains overlap at one
point in their life cycles in that they both inhabit the dog gut. It has
been shown that these two strains have evolved different metabolic pathways, so that competion for the same nutrients is likely to be reduced.

The above hypothesis would appear to be a reasonable explanation for the
appearance of a range of new strains of E.granulosus and E.multilocularis
but it does not exclude the possibility that cross-fertilization may also
occur which could result in genetic recombination of the existing genome
and be a further source of intraspecific variation. It must be pointed out,
however, that although cross-fertilization has been reported in some species of cestodes (12) it has never, to date, been observed in the genus
Echinococcus.

REFERENCES

1. Craig PS, Macpherson CNL and Nelson GS:The identification of eggs of
   Echinococcus by immunofluorescence using a specific anti-oncospheral
   antibody. American Journal of Tropical Medicine and Hygiene, 35: 152-
   158, 1986.
2. Eckert J and Ramp Th: Cryopreservation of Echinococcus multilocularis
   metacestodes and subsequent proliferation in rodents (Meriones).
   Zeitschrift für Parasitenkunde, 71: 777-787, 1985.
3. Jenkins DJ and Rickard MD: Specific antibody response in dogs experimentally infected with Echinococcus granulosus and allowed to continue to
   patency. Coopers Animal Health Symposium on the Immunology and Biology
   of Cestode Infections. University Veterinary School, Melbourne, 1986.
   Abstract E9.

10

4. Krebs CJ: Ecology: The Experimental Analysis of Distribution and Abundance. 2nd Edition. Harper & Row.N.Y. 1978.
5. Kamiya M, Ooi K-H, Oku Y, Yagi K and Ohbayashi M: Growth and development of Echinococcus multilocularis in experimentally infected cats. Japanese Journal of Research, 33: 135-140, 1985.
6. Kumaratilake LM and Thompson RCA: A comparison of Echinococcus granulosus from different geographical areas of Australia using secondary cyst development in mice. International Journal of Parasitology, 13: 509-515, 1983.
7. Kumaratilake LM, Thompson RCA and Dunsmore JD: Comperative strobilar development of Echinococcus granulosus of sheep origin from different geographical areas of Australia in vivo and in vitro. International Journal of Parasitology, 13: 151-156, 1983.
8. Kumaratilake LM, Thompson RCA, Eckert J and Alessandro AD: Sperm transfer in Echinococcus (Cestoda: Taeniidae). Zeitschrift für Parasitenkunde, 72: 265-269, 1986.
9. McManus DP and Smyth JD: Differences in the chemical composition and carbohydrate metabolism of Echinococcus granulosus (horse and sheep strains) and E.multilocularis. Parasitology, 77: 103-109, 1978.
10. McManus DP and Smyth JD: Isoelectric focusing of some enzymes from Echinococcus granulosus (horse and sheep strains) and E.multilocularis. Transactions of the Royal Society of Tropical Medicine and Hygiene, 73: 259-265, 1979.
11. McManus DP and Smyth JD: Intermediate carbohydrate metabolism in protoscoleces of Echinococcus granulosus (horse and sheep strains) and E.multilocularis. Parasitology, 84: 351-366, 1982.
12. McManus DP and Smyth JD: Hydatidosis: changing concepts in epidemiology and speciation. Parasitology Today, 2: 103-168, 1986.
13. Nelson GS: Human behavior in the transmission of parasitic diseases. In: Behavioural Aspects of Parasite Transmission. Canning EU and Wright, CA (Eds) Supplement to Zoo Journal of the Linnean Society, 51:109-122, 1972.
14. Pawlowski ZS: Chemotherapy of human echinococcosis. In: XIII International Congress of Hydatidology, Madrid. April 1985.
15. Reisin IL: (Host-parasite interaction with special reference to the larval stage of Echinococcus granulosus).(In Spanish). In: XIII International Congress of Hydatidology, Madrid, April 1985, 91-95.
16. Reisin IL and Pavisic de Fala CI: Membrane permeability of secondary hydatid cysts of Echinococcus granulosus. Determination of the water diffusional and osmotic permeability coefficients through a syncytial membrane. Molecular and Biochemical Parasitology, 12: 101-116, 1984.
17. Schantz PM: Effective medical treatment for hydatid disease? Journal of the American Medical Association, 253: 2095-2097, 1985.
18. Slepnev NK, Zen'kov AV and Kaminskaya YaM: The susceptibility of farm animals to infection with a strain of Echinococcus granulosus from wild ungulates.Veterinarno Nauka-Proizvodstvu, 21: 91-93,1983.
19. Smyth JD: The biology of the hydatid organisms.Advance in Parasitology, 2: 169-219, 1964a.
20. Smyth JD: Observations on the scolex of Echinococcus granulosus, with special reference to the occurence of secretory cells in the rostellum. Parasitology, 54: 515-526, 1964b.
21. Smyth JD: Strain differences in Echinococcus granulosus with special references to the status of equine hydatidosis in the United Kingdom.

Transactions of the Royal Society of Tropical Medicine and Hygiene, 71: 93-100, 1977.

22. Smyth JD: The insemination-fertilization problem in cestodes cultured in vitro. A Festschrift dedicated to the fiftieth anniversary of the Institute of Parasitology of McGill University 1932-1982. E Meerovitch (Ed.) 393-406,1982. Montreal. McGill University.

23. Smyth JD: The "Niche" concept in Parasitology with special reference to hydatid disease. Proceedings of the British-Scandinavian Joint Meeting in Tropical Medicine and Parasitology. Copenhagen, Sept.1985, p.61.

24. Smyth JD and Barrett NJ: Procedures for testing the viability of human hydatid cysts following surgical removal, especially after chemotherapy. Transactions of the Royal Society of Tropical Medicine and Hygiene, 74: 649-652, 1980.

25. Smyth JD and Davies Z: Occurence of physiological strains of Echinococcus granulosus demonstrated by in vitro culture of protoscoleces from sheep and horse hydatid cysts. International Journal of Parasitology, 4: 443-445, 1974.

26. Smyth JD, Morseth DJ and Smyth MM: Observations on nuclear secretions in the rostellar gland cells of Echinococcus granulosus (Cestoda). The Nucleus, 12: 47-56, 1969.

27. Smyth JD and Smyth MM: Natural and experimental hosts of Echinococcus granulosus and E.multilocularis, with comments on the genetics of speciation in the genus Echinococcus. Parasitology, 54: 493-514, 1964.

28. Smyth JD and Smyth MM: Self-insemination in E.granulosus in vivo. Journal of Helminthology, 43: 383-388, 1969.

29. Swiderski Z: Echinococcus granulosus: hook-muscle systems and cellular organization of infective oncospheres. International Journal of Parasitology, 13: 289-299, 1983.

30. Thompson RCA: (Ed.) The Biology of Echinococcus and Hydatid Disease. George, Allen & Unwin. London. 1986a.

31. Thompson RCA: Biology and systematics of Echinococcus. Chapter 1. In: The Biology of Echinococcus and Hydatid Disease. (Ed. RCA Thompson). George Allen & Unwin. London, 1986b.

32. Thompson RCA, Dunsmore JD and Hayton AR: Echinococcus granulosus: secretory activity of the rostellum of the adult cestode in situ in the dog. Experimental Parasitology, 48: 144-163, 1979.

33. Thompson RCA and Eckert J: Observations on Echinococcus multilocularis in the definitive host. Zeitschrift für Parasitenkunde, 69: 335-345, 1983.

34. Thompson RCA, Kumaratilake LM and Eckert J: Observations on Echinococcus granulosus in Switzerland. International Journal of Parasitology 14: 283-291, 1984.

35. Zhuravets SK: Periods of maturation of Echinococcus granulosus in dogs. Byull.vses. Inst.gel'minth. K.I.Sktyabina, 32:28-32, 1982.

THE FERTILITY OF HYDATID CYSTS IN FOOD ANIMALS IN GREECE

C.HIMONAS, S.FRYDAS AND K.ANTONIADOU-SOTIRIADOU

ABSTRACT
They were examined 601 hydatid cysts from 103 sheep, 194 cysts from 77 goats, 528 cysts from 107 cattle, and 102 cysts from 60 pigs. The results were as follows: a) Sheep: 64,2% of the cysts examined were fertile, 9,1% sterile, and 26,6% calcified. Of the infected animals, 73,8% were carrying fertile cysts, 21,3% sterile, and 27,2% calcified. The cysts were more frequent and numerous in liver, but the fertility rate was higher among the cysts of lungs than in those of liver. b) Goats: 54,6% of the cysts examined were fertile, 34,5% sterile, and 10,8% calcified. Of the infected animals, 57,1% were carrying fertile cysts, 41,5% sterile, and 20,7% calcified. The cysts were more frequent and numerous in lungs. Likewise the fertility rate was higher among the cysts of lungs than in those of liver. c) Cattle: 16,1% of the cysts examined were fertile, 68,9% sterile and 14,9% calcified. Of the infected animals, 30,8% were carrying fertile cysts, 75,7% sterile, and 29,9% calcified. The cysts were more frequent and numerous in lungs, but the fertility rate was higher among the cysts of liver than in those of lungs. d) Pigs: 6,9% of the cysts examined were fertile, 60,8% sterile, and 32,3% calcified. Of the infected animals, 8,3% were carrying fertile cysts, 73,3% sterile and 35% calcified. The cysts were more frequent and numerous in liver. Likewise the fertility rate was higher among the cysts of liver than in those of lungs. No correlation whatewer was found between age of animals and number of cysts harboured.

1. INTRODUCTION

Hydatid cysts in different geographical regions and in different hosts may be of variable fertility (10, 30). This variability may profoundly influence the epidemiology and the control of echinococcosis/hydatidosis (34, 39).

Actually, differences on the fertility rate of hydatid cysts of different animals has been reported by many authors in many countries (1, 2, 3, 4, 6, 8, 9, 13, 14, 16, 17, 18, 20, 21, 22, 23, 24, 25, 29, 30, 35, 37, 40, 41, 46).

In Greece, although the incidence of hydatid infection of food animals has been repeatedly studied in the last 35 years (19, 26, 27, 42, 43, 44), the fertility rate of their hydatid cysts has been never before investigated. For this reason a research survey was undertaken in order to investigate the incidence of fertile cyst and the fertility rate of meat animals slaughtered in the vicinity of Thessaloniki.

2. MATERIALS AND METHODS

During a 14 months period (1984-86) four slaughterhouses in the city and the vicinity of Thessaloniki (Central Macedonia, Northern Greece) were visited regularly and all organs infected with hydatid cysts were collected and brought into the Laboratory.

A total of 103 sheep, 77 goats, 107 cattle and 60 pigs, originating from 9 different localities of Central Macedonia, were found infected. The age of the animals, which was determined by the veterinarians-meat inspectors, varied between 1 to 6 years in sheep, 1 to 3 years in goats, 2 to 10 years in cattle, and 9 months to 8 years in pigs.

They were found 601 ovine hydatid cysts, 194 caprine, 528 bovine and 102 swine. Each cyst, along with some surrounding tissues, was removed from the organ infected and put separately in a petri dish. The hydatid fluid and the germinal membrane of each cyst were examined under a dissecting microscope (X 12,8-102,4) for the presence of any protoscolices. Not any viability test was carried out with the protoscolices found. Also, not any attempt was made to investigate the possible fertility of calcified cysts.

3. RESULTS

The results, which are presented in details on tables 1, 2, 3, 4, 5 and 6, were as follows:

3.1. Sheep: 386 out of 601 hydatid cysts (i.e. 64,2%) were fertile, 55 (i.e. 9,1%) were sterile, and 160 (i.e. 26,6%) were calcified. Seventy-six animals out of the 103 examined (i.e. 73,8%) were carrying fertile cysts, 22 (i.e. 21,3%) sterile, and 28 (i.e. 27,2%) calcified. The number of cysts counted per animal varied between 1 and 52 (M: 5,83 $\pm$ 0,76). The cysts were more frequently found in liver (71 sheep, i.e. 68,9%) and less in lungs (40 sheep, i.e. 38,8%). Eight of these sheep (i.e.7,7%) were hosting hydatid cysts in both liver and lungs. The most numerous cysts (450 i.e.74,8%) were found in liver, while 151 cysts (i.e. 25,1%) were located in lungs. The fertility rate was higher among the cysts of lungs (118 out of 151 cysts, i.e. 78,1%) than in those of liver (268 cysts out of 450, i.e. 59,5%).

3.2. Goats: 106 out of 194 hydatid cysts (i.e. 54,6%) were fertile, 67 (i.e. 34,5%) were sterile, and 21 (i.e. 10,8%) were calcified. Fourty-four animals out of the 77 examined (i.e. 57,1%) were carrying fertile cysts, 32 (i.e. 41,5%) sterile, and 16 (i.e. 20,7%) calcified. The number of cysts counted per animal varied between 1 and 17 (M: 2,52 $\pm$ 0,21). The cysts were more frequently found in lungs (50 goats, i.e. 64,9%) and less in liver (31 goats, i.e. 40,2%). Four out of these goats (i.e. 5,2%) were hosting hydatid cysts in both lungs and liver. The most numerous cysts (122 i.e. 62,8%) were found in lungs, while 72 (i.e. 37,1%) were located in liver. Similarly, the fertility rate was higher among the cysts of lungs (74 out of 122 cysts, i.e. 60,6%) than in those of liver (32 out of 72 cysts, i.e. 44,4%).

3.3. Cattle: 85 out of 528 hydatid cysts (i.e. 16,1%) were fertile, 364 (i.e. 68,9%) were sterile, and 79 (i.e. 14,9%) were calcified. Thirty-three out of the 107 animals examined (i.e. 30,8%) were carrying fertile cysts, 81 (i.e. 75,7%) sterile, and 32 (i.e. 29,9%) calcified. The number of cysts counted per animal varied between 1 and 36 (M: 4,93 $\pm$ 0,53). The cysts were more frequently found in lungs (64 cattle, i.e. 59,8%) and less in liver (46 cattle, i.e. 42,9%). Three out of these cattle (i.e. 2,8%) were hosting hydatid cysts in both lungs and liver. The most numerous cysts (387 i.e. 73,2%) were found in lungs, while 141 (i.e. 26,7%) were located in liver. The fertility rate was higher among the cysts of liver (28 out of 141 cysts, i.e. 19,8%) than in those of lungs (57 out of 387 cysts, i.e. 14,7%).

3.4. Pigs: 7 out of 102 hydatid cysts (i.e. 6,9%) were fertile, 62 (i.e. 60,8%) were sterile, and 33 (i.e. 32,3%) were calcified. Five out of the 60 animals examined (i.e. 8,3%) were carrying fertile cysts, 44 (i.e. 73,3%) sterile, and 21 (i.e. 35%) calcified. The number of cysts counted per animal varied between 1 and 6 (M: 1,7 $\pm$ 0,12). The cysts were more frequently found in liver (36 pigs, i.e. 60%)

TABLE 1. Fertility rate of hydatid cysts of food animals in Greece

| Infected animals | Number of cysts | | Kind of cysts | | | Percentage excluding calcified cysts |
Species-Number	Total	Per animal Variation-Mean	Fertile Number-%	Sterile Number-%	Calcified Number-%	Fertile - Sterile
Sheep - 103	601	1-52 - 5,83	386 - 64,2	55 - 9,1	160 - 26,6	87,5% - 12,5%
Goats - 77	194	1-17 - 2,52	106 - 54,6	67 - 34,5	21 - 10,8	61,3% - 38,7%
Cattle - 107	528	1-36 - 4,93	85 - 16,1	364 - 68,9	79 - 14,9	18,9% - 81,1%
Pigs - 60	102	1- 6 - 1,7	7 - 6,9	62 - 60,8	33 - 32,3	10,1% - 89,9%

TABLE 2. Fertility rate among animals

Animals Species-Number	With fertile cysts Number - %	With sterile cysts Number - %	With calcified cysts Number - %
Sheep - 103	76 - 73,8	22 - 21,3	28 - 27,2
Goats - 77	44 - 57,1	32 - 41,5	16 - 20,7
Cattle - 107	33 - 30,8	81 - 75,7	32 - 29,9
Pigs - 60	5 - 8,3	44 - 73,3	21 - 35

and less in liver (26 pigs, i.e. 43,3%). Two out of these pigs (i.e. 3,3%) were hosting cysts in both liver and lungs, while in one pig (i.e. 1,6%) cysts were found in both lungs and kidneys. The most numerous cysts (60 i.e. 58,8%) were found in liver, 40 cysts (i.e. 39,2%) in lungs and 2 only (i.e. 1,9%) in kidneys. Likewise the fertility rate was higher among the cysts of liver (6 out of 60 cysts, i.e. 10%) than in those of lungs (1 out of 40 cysts i.e. 2,5%). The two cysts found in kidneys were sterile.

Statistical analysis of the data available from sheep and goats did not show any age influence on the number of cysts carried by each age group. Nevertheless a tendency of increase in the number of fertile cysts was noticed with the advancement of age.

4. DISCUSSION

Independently of whether or not the differences on the fertility rate of hydatid cysts, among different intermediate hosts of *Echinococcus granulosus* of different regions, is a matter of intraspecific variation, or of the host-parasite adaptation, or of the local predactor-prey relationship (11, 32), it is a necessity that such differences must be taken into consideration if ecological patterns of transmission in different areas are to be elucidated and effective control measures implemented (38).

Besides other criteria, the accepted intraspecific strains of *E. granulosus* are characterized by a comparative high fertility rate of their hydatid cysts in certain hosts. This way various epidemiological cycles can be determined. For example a horse/dog cycle has been shown to operate in Great Britain (39); a pig/dog cycle seems to exist in U.S.S.R., (36, 48), in Bulgaria (30), in Poland (15), and in Yugoslavia (47); a camel/dog cycle it is known to exist in Africa and Arabia (5, 12, 13, 14, 25, 26); and a cattle/dog cycle has been reported in Sri Lanka (7), South Africa (45), Belgium (6), and Switzerland (8, 9). But the most prevalent epidemiological cycle of *E.granulosus* all over the world is that of sheep/dog. In areas where this cycle predominates, other animals, such as cattle and pigs, can be also infected, but their incidence of infection is low and even more their hydatid cysts are rarely fertile (40).

In Greece, as in most Mediterranean countries, it is generally accepted that sheep operates as the most important intermediate host of *E.granulosus*. This concept, which up to now was based only of the high incidence of its hydatid infection, is now confirmed by the present study, which demonstrated the comparative high fertility rate of hydatid cysts in sheep and goats, and the low one in cattle and pigs. Thus, by adapting the recent data on the hydatid infection

TABLE 3. Classification of the animals according to the kind of hydatid cysts

Animals Species	Total Number	With fertile cysts only Number-%	With sterile cysts only Number-%	With calcified cysts only Number-%	With fertile & sterile cysts Number-%	With fertile & calcified cysts Number-%	With sterile & calcified cysts Number-%	With fertile, sterile & calcified cysts Number-%
Sheep	103	55 - 53,4	6 - 5,8	19 - 18,4	14 - 13,6	7 - 6,8	2 - 1,9	——
Goats	77	31 - 40,2	19 -24,7	12 - 15,6	11 - 14,3	2 - 2,6	2 - 2,6	——
Cattle	107	13 - 12,1	45 -42	13 - 12,1	17 - 15,8	——	16 -14,9	3 - 2,8
Pigs	60	1 - 1,6	37 -61,6	12 - 20	1 - 1,6	3 - 5	6 -10	——

TABLE 4. Location of hydatid cysts

Animals infected	Organs infected			
	Liver	Lungs	Liver & lungs	Lungs & kidneys
Species-Number	Number- %	Number- %	Number- %	Number- %
Sheep - 103	63 - 61,2	32 - 31,1	8 - 7,7	——
Goats - 77	27 - 35,1	46 - 59,7	4 - 5,2	——
Cattle - 107	43 - 40,1	61 - 57	3 - 2,8	——
Pigs - 60	34 - 56,7	23 - 38,3	2 - 3,3	1 - 1,6

and fertility rate percentages onto the actual populations of food animals in Greece, it can easily be determined theoritically the numbers of these animals that may serve as potential sources of dogs' infection, and so to understand the importance of each one of them in the epidemiology of echinococcosis/hydatidosis in this country (Table 7).

5. REFERENCES

1. Andersen FL, Wright PD & Mortenson C: Prevalence of Echinococcus granulosus infection in dogs and sheep in Central Utah. JAVMA 163 (10): 1168-1171, 1973.
2. Attanasio E, Dottorini S & Palmas C: Ecologic and epidemiologic aspects in hydatidosis control programs. XII Congreso Internacional de Hidatilogia, Madrid, Ponencias: 39-46, 1985.
3. Bortoletti G & Fessetti G: Ultrastructural aspects of fertile and sterile cysts of Echinococcus granulosus developed in hosts of different species. Intern.J.Parasit. 8: 421-431, 1978.
4. Cordero del Campillo M: Parasitic zoonoses in Spain. Intern.J.Zoon. 1 (2): 43-57, 1974.
5. Dada BJO, Belino ED, Adegboye DS & Mohammed AN: Experimental transmission of Echinococcus granulosus of "camel-dog" strain to goats, sheep, cattle and donkeys. Intern.J.Zoon. 8: 33-43, 1981.
6. De Rycke PH: Het voorkomen van steriele hydatigen bij runderen en de subspeciatie van Echinococcus granulosus. Vl.Diergen.Tijdschr. 37 (2): 101-108, 1968.
7 Dissanaike AS: Some preliminary observations on Echinococcus infection in local cattle and dogs. Ceylon Med.J. 4: 69-75, 1957. In Thompson et al, 1984.
8. Eckert J: Echinikokkose bei Mensch und Tier.Schweiz.Arch.Tierheilk. 112: 443-457, 1970.
9. Eckert J: Echinokokkose. Berl.Münch.Tierärztl.Woch. 94: 369-378, 1981.
10. Eckert J, Gemmell MA & Soulsby EJL: FAO/UAEP/WHO Guidelines for surveillance, prevention and control of echinococcosis/hydatidosis. Geneva: WHO, VPH/81.28, 1981.
11. Euzeby J: Les echinococcoses animales et leur relations avec les echinococcoses de l' homme. Vigot Fr.Ed., Paris, 1971.
12. Euzeby J: Les echinococcoses larvaires: Presentation du sujet. Sci.Vet.Med.

TABLE 5. Distribution of hydatid cysts in the liver

Animal species	Number of livers	Number of cysts	Kind of cysts					
			Fertile only Nr - %	Sterile only Nr - %	Calcified only Nr - %	Fertile & sterile Nr - %	Fertile & calcified Nr - %	Sterile & calcified Nr - %
Sheep	71	450	205 - 45,5	6 - 1,3	70 - 15,5	44 - 9,8 (f.28 s.16)	105 - 23,3 (f.35 c.70)	20 - 4,4 (s.12 c.8)
Goats	31	72	19 - 26,4	25 - 34,7	9 - 12,5	19 - 26,4 (f.13 s.6)	—	—
Cattle	46	141	18 - 12,8	56 - 39,7	20 - 14,2	20 - 14,2 (f.10 s.10)	—	27 - 19,1 (s.16 c.11)
Pigs	36	60	2 - 3,3	21 - 35	19 - 31,7	—	10 - 16,7 (f.4 c.6)	8 - 13,3 (s.4 c.4)

TABLE 6. Distribution of hydatid cysts in the lungs

Animal species	Number of lungs	Number of cysts	Kind of cysts						
			Fertile only Nr - %	Sterile only Nr - %	Calcified only Nr - %	Fertile & sterile Nr - %	Fertile & calcified Nr - %	Sterile & calcified Nr - %	Fertile, sterile & calcified Nr - %
Sheep	40	151	86 - 56,9	6 - 4	5 - 3,3	40 - 26,5 f.25 s.15	14 - 9,3 f. 7 c. 7	—	—
Goats	50	122	56 - 45,9	25 - 20,5	8 - 6,5	24 - 19,7 f.16 s. 8	4 - 3,3 f. 2 c. 2	5 - 4,1 s. 3 c. 2	—
Cattle	64	387	12 - 3,1	96 - 24,8	10 - 2,6	131 - 33,8 f.42 s.89	—	95 - 24,5 s.60 c.35	43 - 11,1 f. 3 s.37 c. 3
Pigs	26	40	—	31* - 77,5	2 - 5	3 - 7,5 f. 1 s. 2	—	4 - 10 s. 2 c. 2	—

* in one pig (No 15), besides 1 sterile cyst in the lung, 2 sterile cysts were found in one kidney

20

TABLE 7. Potential number of livestock infected with fertile hydatid cysts in Greece (data of 1984[1], 1985[2], 1986[3])

Livestock		Hydatid infection		Fertility rate	
Species	Population[1]	Incidence[2]	Number of animals infected	Incidence[3]	Number of animals infected*
Sheep	8.500.000	X 51,16%	= 4.348.600	X 73,78%	= 3.208.397
Goats	4.650.000	13,50%	627.750	57,14%	358.696
Cattle	800.000	33,31%	266.480	30,84%	82.182
Pigs	1.324.000	4,65%	61.566	8,33%	5.128

* potential sources of dogs' infection

Comp. 85 (2): 67-78, 1983.
13. Graber M, Troncy P, Tabo R, Service J & Oumatie O: L'echinococcose-hydatidose en Afrique centrale. I.Echinococcoses des animaux domestiques et sauvages. Rev.Elev.Med.Vet.Pays Trop. 22 (1): 55-67, 1969.
14. Islam AWKS: Hydatidosis in sheep in Bangladesh. Vet.Med.Rev. 2/81: 151-152, 1981.
15. Kozakiewitz B: Invasiveness of Echinococcus granulosus cysts in pigs. XI Intern.Congr.Hydatidosis. Athens: 49, 1977.
16. Le Riche PD, Dwinger RH & Kühne GI: Bovine echinococcosis in the North-West Argentina. Trop.Anim.Hlth Prod. 14: 205-206, 1982.
17. Lubke R: Betrachtungen über die Fertilität der Sweineechinokokken und ihr Verhätnis zur Zystendegeneration durch die Abwehrreaktion des Organisms. Tierärztl.Umsch. 5: 1-8, 1972.
18. Luttermoser GW & Koussa M: Epidemiology of echinococcosis in the Middle East. II.Incidence of hydatid infection in swine in Lebanon. Amer.J.Trop.Med.& Hyg. 12, 22-25, 1963.
19. Maccas M: Statistique de l'echinococcose hymaine en Grèce. Arch.Intern.Hidatid. 12 (1-2): 61-70, 1951.
20. Macchioni G & Gallo C: Sulla presenza in Italia dell' Echinococcus granulosus equinus Williams e Sweatman. Ann.Fac.Med.Vet., 22: 58-77, 1967.
21. Martinez Gomez F, Hernandez Rodriguez S, Navarette Lopez-Cozar I & Calero Carretero R: Serological tests in relation to the viability, fertility, and localizations of hydatid cysts in cattle, sheep, goats and swine. Vet.Parasit. 7: 33-38, 1980.
22. Mendy RM: Hydatid zoonosis. Vet.Med.Rev. 1/2: 176-191, 1975.
23. Mirzayans A: The incidence of hydatidosis and other cestode larvae in domestic animals in Tehran abattoir. J.Vet.Fac.Tehran 30 (4): 14, 1975.
24. Pandey VS: Hydatidosis in donkeys in Marocco. Ann.Trop.Med.Parasit. 74 (5): 519-521, 1980.
25. Pandey VS, Oudelli H & Duchton M: Hydatidosis/echinococcosis in animals in Marocco. XII Congreso Internacional de Hidatidologia, Algiers. Abstracts No 38: 40, 1981.
26. Papachristophilou Ph: L' hydatidose en Grèce chez les ruminants et les porcs. Bull.Off.Intern.Epiz. 47: 469-485, 1957.

27. Papadopoulos G (Mediterranean Zoonosis Control Center): Personal communication.

28. Pipkin AS, Rizk E & Balikian J: Echinococcosis in the Near East and its incidence in animals hosts. Trans.R.Soc.Trop.Med.& Hyg. 45: 253-260, 1951. In Troncy, 1968.

29. Polydorou K: Epidemiology of hydatid cysts and socio-economic consequences with a special reference to Cyprus. XII Congress Internacional de Hidatidologia. Abstracts No 42: 45-46, 1981.

30. Popov A, Bankov D, Momov M, Djankov I, Depev J, Georgiev B, Trifonov T, Angelov G, Bratanov V & Yachkov TS: A study of echinococcosis in Bulgaria. Vet.Sci. (Sofia) 6: 71-81, 1968.

31. Prasat BN & Mandal LV: Incidence of hydatid cysts in buffaloes in Bihar. Kerala J.Vet.Sci. 10 (2): 220-225, 1979. In Vet.Bull. 51 (4) No 1760: 229, 1981.

32. Schantz PM: Echinococcosis. In CRC Handbook Series in Zoonoses. Section C: Parasitic Zoonoses, vol. 2 pp: 231-277 (ed. P.Arambullo), 1982.

33. Schwabe CW: Epidemiology of echinococcosis. Bull. WHO, 39: 131-135, 1968.

34. Smyth JD: Strain differences in Echinococcus granulosus with special reference to the status of equine hydatidosis in the United Kingdom. Trans.R.Soc. Trop.Med.& Hyg. 71: 93-100, 1977.

35. Soulsby EJL: Helminths, Arthropods & Protozoa of Domesticated Animals (Six Edition of Mönnig's Veterinary Helminthology and Entomology). Baillere, Tindal & Cassell, London, 1968.

36. Stoimenov K & Kaloyanov ZH (in Russian). In Thompson et al, 1984.

37. Thompson RCA: Hydatidosis in Great Britain. Helminth.Abstracts Ser A, 46 (10): 837-861, 1977.

38. Thompson RCA: Intraspecific variation and parasite epidemiology. In "Parasites- Their World and Ours" pp 369-378 (ed.Mettrick DP & Desser SS) Elsevier Biomedical Press, Amsterdam, 1982.

39. Thompson RCA & Smyth JD: Equine hydatidosis: a review of the current status in Great Britain and the results of an epidemiological survey. Vet.Parasit. 1: 107-127, 1975.

40. Thompson RCA, Kumaratilake LM & Eckert J: Observations on Echinococcus granulosus of cattle origin in Switzerland. Intern.J.Parasit. 14 (3): 283-291, 1984.

41. Troncy PM: Echinococcose-Hydatidose dans le Bassin Tschadien. These pp 157, Alfort, 1968.

42. Vassalos M: Echinococcosis-Hydatidosis, pp 107, Athens, 1966.

43. Vassalos M, Saravanos A & Tsaglis A: The frequency of hydatidosis of ruminants and pigs in Greece. XI Intern.Congr.Hydatid., Athens, Abstracts: 103-104, 1977.

44. Vassalos M, Himonas C & Saravanos A: Hydatidosis in Greece. In "Some important parasitic infections in bovines, considered from the economic and social (zoonosis) points of view", pp. 205-213 (ed.J.Euzeby & J.Gevrey). Parasitological Symposium, Lyon, 1983. CEC Brussels & Luxenburg, 1984.

45. Verster A: Hydatidosis in the republic of South Africa. South Afr.J.Sci. 58: 71-74, 1962. In Thompson et al, 1984.

46. Visnjakov J & Dimitrov D: Die Verbreitung der Echinokokkose bei Schafen und Ziegen in der Volksrepublik Bulgarien. Wien. Tierärztl.Monatsc. 56 (9): 384-386, 1969.

47. Wikerhauses T & Brglez J: An attemp to infect calves with Echinococcus granulosus of porcine origin. EMOP IV Abstracts No 135 (FP): 122, 1984.

48. Zenkov AV (in Russian): In Thompson et al, 1984.

A FAILURE TO INFECT DOGS WITH <u>ECHINOCOCCUS GRANULOSUS</u> PROTOSCOLECES OF
HUMAN ORIGIN

T.WIKERHAUSER, J.BRGLEZ, M.ŠTULHOFER and B.IVANIŠEVIĆ

1. ABSTRACT

Two large hepatic cysts of Echinococcus <u>granulosus</u>, with numerous fertile daughter cysts, were surgically removed from two female patients, 70
and 36 year old, respectively. The protoscoleces were collected and microscopically checked for shape and motility. The material from each cyst was
administered orally to 3 young dogs. After 3 months the dogs were sacrificed and examined. In none of them either mature or immature E.granulosus
tapeworms were found.

2. INTRODUCTION

In a previous study (2) we have found that bovines and ovines are not
normally susceptible to the infection with Echinococcus <u>granulosus</u> of porcine origin. In the present work we tried to isolate human strains of
E.<u>granulosus</u> and to study their infectivity for pigs, sheep and cattle. The
results were expected to provide information about the possible origin of
infections in humans.

3. MATERIALS AND METHODS

Two large hepatic cysts of E.<u>granulosus</u>, containing numerous fertile
daughter cysts, were surgically removed from two different female patients,
70 and 36 year old, respectively. Both patients came from sheep-rising regions, Dalmatia and Bosnia, respectively, where hydatid disease is common
in man and animals, especially sheep. The protoscoleces were collected and
microscopically checked for shape and ability to devaginate. In the aliquot
sample from the older patient about 5%, and in that of the younger one,40%
of protoscoleces were alive. Within 24 hours after collection several
thousands of protoscoleces from each cyst were administered orally to
three 6-week old, male, cross-bred, puppies. The dogs had no previous contact with hydatids. Periodical fecal examinations were negative. Three
months later they were sacrificed and examined for E.<u>granulosus</u> tapeworms.

4. RESULTS AND DISCUSSION

In none of the dogs, which received protoscoleces of human origin,
E.<u>granulosus</u> were found.

In most parts of the world where hydatid disease occurs in man and
animals the former host is not involved in the transmission of the parasite
to the definitive host(s). Thus, the metacestode in an "abnormal" intermediate host may lose its full capacity for procreation. On the other
hand, when natural transmission from man to dog <u>is</u> likely to occur, as is
the case in Turkana (Kenya), an experimental infection of dogs with protoscoleces of human origin was readily achieved (1).

REFERENCES

1. Macpherson CNL, Karstad L, Stevenson P and Arundel JH: Hydatid disease in the Turkana District of Kenya. III. The significance of wild animals in the transmission of Echinococcus granulosus, with particular reference to Turkana and Masailand in Kenya. Annals of Tropical Medicine and Parasitology, 77: 61-73, 1982.
2. Wikerhauser T, Brglez J and Kutičić V: Eksperimentalno istraživanje otpornosti teladi i janjadi na Echinococcus granulosus svinjskog porijekla. Veterinarski Arhiv, 56: 7-11, 1986.

EXPERIMENTAL INFECTION OF SHEEP AND MONKEYS WITH THE CAMEL STRAIN
OF ECHINOCOCCUS GRANULOSUS

G.MACCHIONI, M.ARISPICI, P.LANFRANCHI, F.TESTI.

ABSTRACT
    In Somalia hydatid cysts are frequently observed in camels (14.82%),
seldom in cattle (1.75%) and exceptionally in goats and sheep; a high
frequency of Echinococcus granulosus infection is found in the stray dogs
in Mogadishu (23.4%) and the parasite is present in the jackals (Canis
mesomelas)too. Notwithstanding the remarkable rate of infection in camels
and the favourable chances to contract the disease, human hydatidosis has
never been noted in Somalia.
    There is evidence from analysis and epidemiological studies that the
camel form of the parasite from Somalia may represent a new strain of
E.granulosus.
    Experimental infection of sheep with the camel strain has confirmed
the low pathogenicity for this animal species; a year after infection the
sheep showed small and immature hydatid cysts in liver and lungs.
    In experimentally infected monkeys (Cercopithecus aethiops) immature
hydatid cysts were present in lungs, liver, kidneys, heart and peritoneum.

INTRODUCTION
    Epidemiological studies on echinococcosis-hydatidosis in Somalia have
shown the occurrence of the cystic stage of Echinococcus granulosus mainly
in camels (14.82%), rarely in cattle (1.75%) and exceptionally in sheep
and goats; the adult tapeworms have been frequently found in stray dogs
(23.4%) and also in jackals (Canis mesomelas)(4).
    It is particularly remarkable that, apart from camels, the infection
is uncommon in those host species which in other parts of the world are
the usual intermediate host, such as sheep and cattle.
    In contrast to the high percentage of infected dogs, human hydatidosis
has never been found in Somalia. The public health services existing in
Somalia during the last about 20 years may have been inefficient, in recent years there has been a considerable improvement in the sanitary situation; thanks to the work of a number of Italian experts in technical
cooperation.
    There is a problem in interpreting the role of the dog-camel cycle in
human hydatidosis in Africa, because other host like cattle, sheep, goats
and wild ungulates also remain exposed to the risk of infection at the
same time.
    The various techniques used to distinguish the strains of E.granulosus
in Kenya have shown a close affinity between the human and the ovine form.
The situation regarding data from camel hydatid remains puzzling in that
the strain responsible for the hydatidosis in this host differes from the

human form both with regards to isoenzyme profiles and basic biochemistry (5, 6).

The status of E.granulosus from camels being unclear, particularly with regard to its potential infectivity to man, it is extremely important to study the Somali camel strain as the available evidence suggest that there is only one strain, with particular behavioural characteristics, which exists in this country.

In an attempt to throw som light on the problem, we have studied the infectivity of the camel strain of E.granulosus to nonhuman primates and sheep.

MATERIALS AND METHODS

Five puppies were each fed a gelatine capsule containing 0.4ml packed protoscoleces of camel origin and were killed 50 days after the oral inoculation. An average of 3,000 to 5,000 adults of E.granulosus was found in the intestine of all the dogs.

Three monkeys (Cercopithecus aethiops) and ten black head sheep each received approximately 10 gravid proglottids orally; two monkeys and three sheep were kept as uninfected control.

Of the three infected monkeys one died one month after infection as a result of septicaemia due to necrotic lesions on its tail. The other died one year post infection after suffering from severe dyspnoea and dry cough; at the same time all the infected and control animals were killed excepting three infected sheep which were pregnant; these were later used to check the subsequent level of development of the cysts.

RESULTS

The examination of the monkey viscera one year after infection revealed the lungs spotted with transparent cysts which ranged in size from 5 to 10 mm in diameter and lacked internal protoscoleces. Similar cysts were observed in the liver and one each in the kidneys, heart and peritoneum of the dead monkey and one cyst in the kidneys of the killed monkey.

The cysts were not histologically identical but often differed from each other in the thickness and staining affinity of the laminated membrane and in the growth and composition of the adventitial reaction.

The germinal layer normally consisted of a thin stratum of cells, sometimes with varying in size from 8 to 24 μm: possibly brood capsules in the intitial stage of development.

The necropsy of the sheep showed that all seven animals had numerous pulmonary and hepatic cysts. They were transparent with a maximum of 5 mm diameter and lacked internal protoscoleces. The histological control confirmed the parasitic origin of the cysts with well formed laminated, germinal and adventitial membranes. In some cysts globose buds varying from 10 to 28 μm in diameter were clearly seen: some had compact aspects, while others contained a fairly large cavity; possibly developing brood capsules. In many cases the cysts were so close to one another as to acquire a multilocular aspect.

CONCLUSIONS

Dada et al. (1) observed in Nigeria that the camel strain of E.granulosus was infective for goats, sheep and, to a lesser extent, cattle; it was not infective for donkeys. Sheep, when necropsied six months after

26

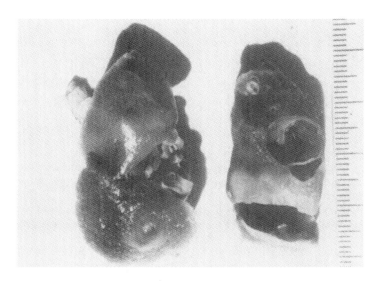

FIGURE 1. Lungs of _Cercopithecus aethiops_
one year after the experimental infection
with camel strain of E. _granulosus_.

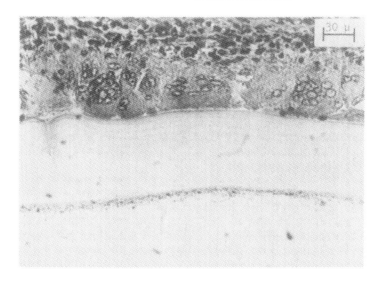

30 μ

FIGURE 2. Liver of _Cercopithecus aethiops_.
Adventitial, laminated and germinal layers
of hydatid cyst.

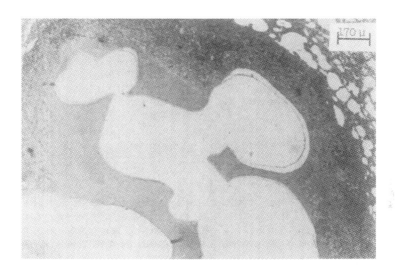

FIGURE 3. Lung of black head sheep. Multilocular aspect of hydatid cysts.

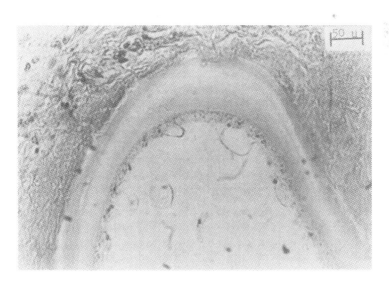

FIGURE 4. Lung of black head sheep. Developing brood capsules in hydatid cyst.

28

infection, had multiple hydatid cysts, measuring 4-5 mm in diameter, in their lungs, liver and spleen.

Our results in Somalia confirm that the camel strain is infective for sheep; one year after infection, the hydatid cysts are small in size (maximum of 5 mm diameter), transparent and sterile. Apart from the variable conditions in the experimental infections of Nigerian and Somali sheep, the same size of the cysts six months and one year after infection suggests a slow capacity of development of the camel strain in this animal species. According to Heath (3) E.granulosus cysts increase in diameter by between 10 and 50 mm per year. It was not possible to know if the cysts in sheep could develop further with protoscoleces formation because one year can be insufficient for complete development of hydatids; in fact, Heath (3) observed that the time of brood capsule growth is extremely variable in sheep, ranging from ten months to four years, and Gemmel and Lawson (2) found protoscoleces in cysts two years after infection. We are waiting for final data from three other experimentally infected sheep to enable us to check the development of hydatid cysts over a long period of time.

In experimental infection of C.aethiops had a positive result and hydatid cysts were found in various organs such as the lungs, the liver, the heart, the kidneys and the peritoneal cavity. In the lungs, the cysts were so crowded together as to cause the death of one monkey with respiratory distress.

It remains difficult to explain the epidemiological data regarding hydatidosis in man in Somalia in the light of what we have observed in monkeys. These results prompt us to carry out further experimental infections on other species of primates in order to check the pathogenicity of the Somali strain of E.granulosus and to undertake a more detailed epidemiological study of the Somalia situation of hydatidosis.

REFERENCES
1. Dada BJO, Belino ED, Adegboye DS, Mohammed AN: Experimental transmission of Echinococcus granulosus of "camel-dog" origin to goats, sheep, cattle and donkeys.International Journal of Zoonoses, 8: 33-43, 1981.
2. Gemmell MA, Lawson JR: Epidemiology and control of hydatid disease. In: The Biology of Echinococcus and Hydatid Diseases, R.C.A. Thompson (ed): 189-216. London: George Allen & Unwin Publ, 1986.
3. Heath DD: The life cycle of Echinococcus granulosus. In: Recent Advances in Hydatid Diseases, R.W.Brown, J.R.Salisbury, W.E.White (eds): 7-18. Victoria, Australia: Hamilton Medical and Vterinary Association, 1973.
4. Macchioni G: Epizootiology and epidemiology of hydatidosis-echinococcosis in Somalia. 6th Symposium on Problems in Tropical Veterinary Medicine, Liblice. Czechoslovakia, May 29-31 (in press), 1985 .
5. McManus DP: A biochemical study of adult stages of Echinococcus granulosus of human and animal origin from Kenya. Journal of Helminthology, 55: 2-27, 1981.
6. Macpherson CNL, McManus DP: A comparative study of Echinococcus granulosus from human and animal hosts in Kenya using isoelectric focusing and isoenzyme analysis. International Journal of Parasitology, 12: 515-521, 1982.

CHARACTERISATION OF THE HYDATID DISEASE ORGANISM, ECHINOCOCCUS
GRANULOSUS, FROM KENYA USING CLONED DNA MARKERS.

D.P. McManus[1], A.J.G. Simpson[2] and A.K. Rishi[1]

ABSTRACT
        Cloned DNA fragments of the ribosomal RNA gene of Schistosoma
mansoni hybridised strongly to DNA of isolates of Echinococcus granulosus
obtained from Kenya, following restriction endonuclease and Southern
transfer analysis. The hybridisation patterns were similar for E.
granulosus of sheep, human, goat and cattle origin but substantial DNA
differences were obtained between these isolates and those of camel
origin. We conclude that the camel cystic material is very distinct
genetically from the other isolates of E. granulosus examined in this
study. This corroborates previous biochemical and isoenzyme data and
suggests that at least two strains of E. granulosus occur in the Turkana
district of Kenya.

INTRODUCTION
        It is now recognised that biologically distinct sub-specific
variants or strains of Echinococcus granulosus, the causative agent of
unilocular hydatid disease, occur, and that these may vary in their
infectivity to domestic animals and probably to man (reviewed, Reference
1). Classical morphological, in vitro culture and biochemical methods such
as protein and enzyme (zymogram) analysis have been applied earlier for
characterisation of E. granulosus isolates from several endemic areas,
including the U.K., Australia and Kenya (1-3). However, a molecular
approach involving restriction-enzyme analysis and Southern blot
hybridisation provides greater specificity because the technique can detect
minute differences in DNA sequences between very closely related organisms.
DNA analysis has the additional advantage over other identification methods
in that individual genes or parts of genes are studied rather than their
expressed products and the approach is unlikely to show any life-stage or
environmentally mediated, including host induced, variation. Recently,
cloned DNA fragments of the ribosomal RNA gene family of Schistosoma
mansoni have been used as genetic markers to discriminate between isolates
of E. granulosus from U.K. horses and sheep (4). Here we describe the
results of applying this approach for characterisation of Kenyan isolates
of E. granulosus. In Kenya there is a formidable public health problem
involving human hydatidosis in the Turkana district (5) and a wide range of
animal hosts, including goats, sheep, camels and cattle, harbour E.
granulosus there.

MATERIALS AND METHODS
Materials
        Restriction endonucleases were purchased from Pharmacia Biochemicals
Ltd. Digestions were carried out in excess using the conditions recommended

by the manufacturers. Hgt agarose was obtained from Seakem. $(\alpha-^{32}P)dCTP$ was purchased from Amersham International. All other chemicals were of analytical grade and were obtained from the Sigma Chemical Corporation or British Drug Houses. All glassware was autoclaved or millipore filtered (Millipore Corporation, 0.45 um pore size) prior to use.

Preparation of DNA

DNA was isolated from the following:-
A). Protoscoleces of E. granulosus of human, sheep, camel, goat and cattle origin obtained from hydatid cysts at surgery or from abattoirs in Kenya. Cysts of cattle origin were obtained from the Central Abattoir of the Kenya Meat Commission, Athi River, Nairobi, those of sheep and goat origin from Ongata Rongai Slaughterhouse, Nairobi and camel material from a rural abattoir in Lodwar, Turkana District. Human cysts were surgically removed from patients at Lodwar Hospital. Protoscoleces were obtained from cysts by treatment in Hanks' saline, pH 2.0, containing 0.2% (w/v) pepsin, followed by several rinses in Hanks' saline (6).
B). Liver tissue of goat, sheep and camel origin.

Parasites and liver tissues were stored in liquid nitrogen, or at $-70^{\circ}C$ until required for DNA extraction. Total DNA was isolated using modifications of a procedure previously described (7). Frozen material was crushed in liquid nitrogen and thawed into 4ml of extraction buffer (50mM Tris-HCl, pH 8.0; 0.5 mM EDTA; 100mM NaCl). To the suspension was added an equal volume of extraction buffer containing 1% (w/v) SDS and 1mg proteinase K. The solution was incubated at $37^{\circ}C$ for 3h with intermittent, gentle shaking. The total nucleic acids, thus released, were extracted with an equal volume of buffer-equilibrated phenol, followed by phenol-chloroform and chloroform extractions and finally precipitated in 300mM Na acetate with 2.5 volumes of cold absolute alcohol overnight at $-20^{\circ}C$. The precipitated nucleic acids were pelleted at 10,000g for 10 min at $0^{\circ}C$, dried and dissolved in 3-5 ml 10mM Hepes, pH 7.5. With some samples, 3 vol of 4M LiCl was added, the solution mixed and left overnight at $4^{\circ}C$. RNA was pelleted at 10,000g for 10 min at $4^{\circ}C$ and DNA was isolated from the supernatent by addition of 2.5 vol of cold absolute alcohol and subsequent spooling. Alternatively, total nucleic acid fractions were subjected to treatment with DNAse-free RNAse, extracted with phenol, phenol-chloroform and chloroform as above and DNA isolated by spooling, following addition of Na acetate and absolute alcohol. Spooled DNA was washed in 70% alcohol, dried and dissolved in 0.5ml T.E. buffer (10mM Tris-HCl, pH 8.0; 1mM EDTA).

Calf-thymus DNA and human placental DNA were obtained commercially.

Preparation of recombinant plasmids, pSM889 and pSM890

Escherichia coli HB101 cells, harbouring recombinant plasmids, pSM889 and pSM890 (8) were grown in 1 litre of LB medium and amplified overnight in the presence of chloramphenicol. Colonies were harvested as pellets and lysed in 0.2M NaOH; 1.0% SDS; DNAse-free RNAse. DNA was precipitated with 750mM K acetate and 0.6 vol isopropanol at $-20^{\circ}C$. DNA pellets were dried, suspended in T.E. buffer and centrifuged at 45,000 rpm (Sorvall 50 VTi rotor) for 18h at $20^{\circ}C$ in the presence of CsCl and ethidium bromide. The plasmid DNA bands were extracted with isopropanol, dialysed against T.E. buffer, precipitated at $-20^{\circ}C$, pelleted by centrifugation, washed in 70% ethanol, dried and dissolved in T.E. buffer.

Hybridisation of (α^{32}P)-labelled plasmid DNAs to restriction fragments of E. granulosus and mammalian DNAs

DNA samples of E. granulosus and mammalian origin were individually digested with the restriction enzymes, EcoR1 and BamH1. The DNA restriction fragments were then electrophoresed on 0.8% agarose gel, transferred to nitrocellulose filters according to the method of Southern (9). The filters were then subjected to hybridisation in 30 ml prehybridisation solution (4 x Denhardt's solution (10); 4 x 0.15 M NaCl/0.015 M Na-citrate (SSC); 0.1% SDS and 100 μg/ml sheared and denatured salmon sperm DNA) in the presence of (α^{32}P)-labelled DNA (circa 1 ug, labelled with (α^{32}P)dCTP by nick translation (11) from each of the plasmids. The hybridisation was carried out at 65°C for 12-16 h. The filters were washed for three hours in 0.1 x SSC; 0.05% SDS at 52°C with four changes of 45 min each, dried and set up for autoradiography for 3-5 days using Kodak X-ray film backed by double cronex lighting plus intensifying screens.

RESULTS AND DISCUSSION

Cloned S. mansoni rDNA genes, pSM890 (Figure 1a,b) and pSM889 (Figures 2a,b; 3a,b) hybridised strongly to DNA extracted from protoscoleces of all tested Kenyan isolates of E. granulosus following restriction endonuclease and Southern transfer analysis. The hybridisation patterns were similar for E. granulosus DNAs of sheep, human, goat and cattle origin; in contrast substantial differences in the hybridisation patterns were obtained between these and DNA isolated from E. granulosus of camel origin. The rDNA genes hybridised weakly or not at all to the DNA restriction fragments of mammalian origin and the hybridisation patterns did not interfere with the interpretation of those obtained with the E. granulosus DNA.

We conclude that the camel material is very distinct genetically from the other isolates examined in this investigation. This supports earlier biochemical studies which showed that Kenyan hydatid material of human, sheep, cattle and goat (common variety) origin differed only slightly in isoenzyme profiles and basic biochemistry whereas all camel material and some goat (rare variety) isolates were shown by isoenzyme analysis to be quite distinct (12). This accumulated data suggests that at least two strains of E. granulosus occur in Turkana; a sheep/dog strain, probably similar to that present in other endemic areas, which can cycle through goats, cattle and uniquely through man (13) and a separate camel/dog strain which may cycle through goats. This is not to say that the camel/dog strain is non-infective to man in Turkana but that, to date, all the human hydatid material that we have examined (i.e. protoscoleces obtained from fertile cysts at operation) from this area conforms biochemically and by DNA analysis to the characteristics of the sheep/dog form. Currently, the basis of our identification procedures relies on the availability of protoscoleces and, consequently, we have not carried out any analytical work on poorly developed, dead or sterile human cysts. It is possible that such cysts could arise as a result of patients ingesting eggs of dog/camel origin.

Recent in vitro culture studies by Macpherson & Smyth (14) have shown that Kenyan hydatid material of camel, cattle, goat, human and sheep origin appear to possess similar, if not identical, nutritional and/or physiological requirements. Moreover, it has been shown that eggs of E. granulosus of Kenyan dog/camel origin will establish in the baboon (Papio

Figure 1. ^{32}P labelled rDNA of Schistosoma mansoni (pSM890) hybridised to EcoR1-derived (panel a) and BamH1-derived (panel b) fragments of Echinococcus granulosus (human, sheep, goat, camel and cattle origin from Kenya) and mammalian DNAs.
The number of E. granulosus isolates tested, although not all are shown, was as follows:- human, 7; sheep, 12; cattle, 6; goat, 8; camel, 2.

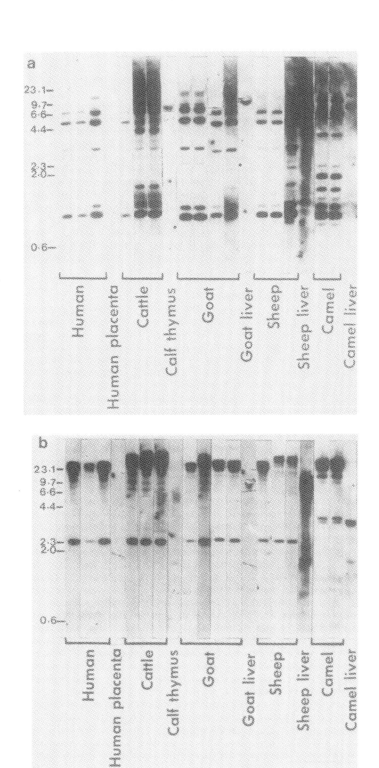

33

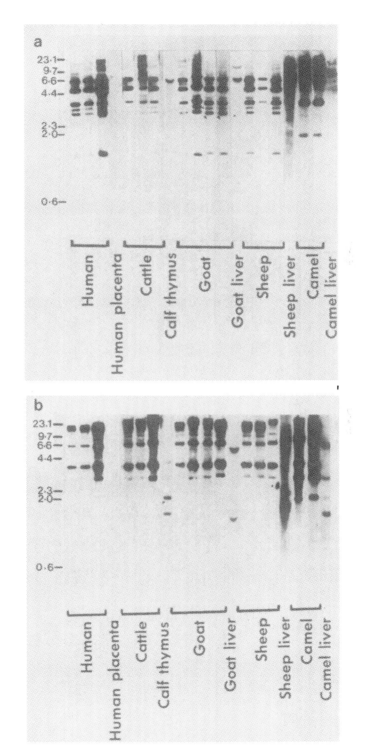

Figure 2. ^{32}P labelled rDNA of Schistosoma mansoni (pSM889) hybridised to EcoR1-derived (panel a) and BamH1-derived (panel b) fragments of Echinococcus granulosus (human, sheep, goat, camel and cattle origin from Kenya) and mammalian DNAs.
The number of E. granulosus isolates tested was as in Figure 1.

34

Figure 3. ³²P labelled rDNA of Schistosoma mansoni (pSM889) hybridised to EcoR1-derived (panel a) and BamH1-derived (panel b) fragments of Echinococcus granulosus (human, sheep, goat, camel and cattle origin from Kenya) DNAs.

The number of E. granulosus isolates tested was as in Figure 1. Note that the hybridisation patterns (especially the smallest fragment) for the goat and cattle isolates shown in panel a are slightly out of alignment with those of the sheep and human isolates. This was due to the fact that the samples were electrophoresed on separate gels but to confirm the similarity of the hybridisation patterns, refer to panel a of Figure 2.

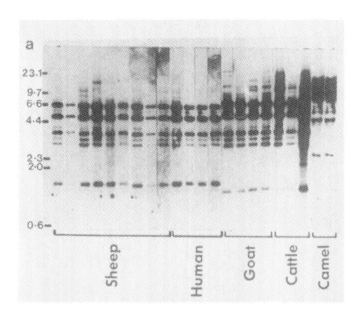

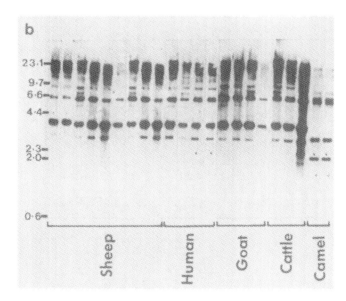

<u>cyanocephalus</u>) (15) although significantly, the cysts obtained following autopsy were either calcified or sterile. In earlier studies, Dailey & Sweatman (16) successfully cross-infected sheep, camels and long-tailed macaque monkeys (<u>Macaca irus</u>) with camel/dog and cattle/dog eggs and concluded that sheep, camels and cattle in the Lebanon are all infected with the same parasite, which they named <u>E. granulosus granulosus</u>. On the other hand, Dada, Belino, Adegboye & Mohammed (17) found that eggs of camel/dog origin from Nigeria developed poorly in goats and sheep, less well in cattle and not at all in donkeys. Moreover, Macchioni (18) has reported some very interesting, recent epidemiological findings concerning hydatidosis in Somalia, where almost a third of the world's camel population is concentrated. The prevalence of <u>E. granulosus</u> in camels was 14.8%, whereas it was lower in cattle (1.75%), sigificantly lower in sheep (0.08%) and not present in goats. At Mogadiscio, one of the areas surveyed by Macchioni, 23.4% of dogs harboured adult <u>E. granulosus</u> and it would have been anticipated that a high human incidence would occur in man in Somalia. However, apart from a single arbitary hydatid in a Somalian woman, human infection has never been recorded in Somalia. Although circumstantial, this evidence strongly suggests that the camel/dog strain from Somalia does not infect man. Isoenzyme analysis of <u>E. granulosus</u> from Somali camels and dogs has, additionally, shown differences between this material and human and sheep <u>E. granulosus</u> of Italian origin (18). It is not known whether the Somalian strain is similar to the one that infects Kenyan camels but it is reasonable to conclude that it is because the two are neighbouring countries and the Somalian nomads often drive their herds across the Kenyan border in search of better pastures and more favourable markets.

Clearly more DNA analysis, of the type employed here, is required to investigate camel hydatid isolates from geographical locations other than Kenya. This may throw light on whether different strains of <u>E. granulosus</u> exist in camels in different areas and may pinpoint which of these strains, if any, are infective to man.

ACKNOWLEDGEMENTS

DPM would like to thank the MRC, Wellcome Trust and EEC for financial support. We are indebted to Calum Macpherson, Caroline Macpherson and Thomas Romig of the African Medical and Research Foundation (AMREF) for supplying some of the parasite material used in this investigation.

REFERENCES
1. McManus D.P. and Smyth J.D., Parasitology Today, <u>2</u>, 163-168, (1986).
2. McManus D.P. and Macpherson C.N.L., Ann. Trop. Med. Parasitol., <u>78</u>, (1984).
3. Thompson R.C.A., Biology and Systematics of <u>Echinococcus</u>, In The Biology of <u>Echinococcus</u>, Ed. Thompson R.C.A., 5-43, George Allen and Unwin, London, (1986).
4. McManus D.P. and Simpson A.J.G., Mol. Biochem. Parasitol., <u>17</u>, 171-178, (1985).
5. Nelson G.S., Trans. Roy. Soc. Trop. Med. Hyg., <u>80</u>, 177-182, (1986).
6. McManus D.P. and Smyth J.D., Parasitology, <u>77</u>, 103-109, (1978).
7. McManus D.P., Knight, M. and Simpson, A.J.G., Mol. Biochem. Parasitol., <u>16</u>, 251-266, (1985).
8. Simpson A.J.G., Dame J.B., Lewis F.A. and McCutchan T.F., Eur. J. Biochem., <u>139</u>, 41-45, (1984).
9. Southern E.M., J.Mol.Biol., <u>98</u>, 503-517, (1975).

10. Denhardt D.T., Biochem. Biophys. Res. Commun., 23, 641-646, (1966).
11. Rigby P.W.J, Dieckmann M., Rhodes C. and Berg P., J. Mol. Biol., 113, 237-251, (1977).
12. Macpherson C.N.L. and McManus D.P., Int. J. Parasitol., 12, 515-521 (1982).
13. Macpherson C.N.L., Am. J. Trop. Med. Hyg., 32, 397-404, (1983).
14. Macpherson C.N.L. and Smyth J.D., Int. J. Parasitol., 15, 137-140, (1985).
15. Macpherson C.N.L., Else J.E. and Suleman M., J. Helminthol., 60, 213-217, (1986).
16. Dailey M.D. and Sweatman G.K., Ann. Trop. Med. Parasitol., 59, 463-477, (1965).
17. Dada B.J.O., Belino E.D., Adegboye D.S. and Mohammed A.H., Int. J. Zoonoses, 8, 33-43 (1981).
18. Macchioni G., Epizootiology and epidemiology of hydatidosis-echinococcosis in Somalia, In the Sixth Symposium on Problems in Tropical Veterinary Medicine, Liblice, Czechoslovakia, 29-31 May, 1985, (in press).

"Kinetics of molecular transfer across the tegument of protoscoleces and hydatid cysts of Echinococcus granulosus and the relevance of these studies to drug targeting"

JEFFS, S.A., HURD, H., ALLEN, J.T. & ARME, C.

ABSTRACT

Protoscoleces of Echinococcus granulosus (horse strain) were found to absorb, by a combination of mediated transport and passive diffusion, 10 L-amino acids, 1 D-amino acid (D-alanine) and D(+)glucose. The L-amino acids enter the protoscolex through four, kinetically distinct, transport systems. Secondary hydatid cysts were also found to absorb and accumulate 4 amino acids, but at a far slower rate than the protoscoleces. Entry into the cyst wall component appears to occur by passive diffusion but, for L-threonine at least, entry into the fluid-filled interior is by mediated transport. A synthetic polymer (PVP) was found to enter the cyst by endocytosis. Analysis of hydatid fluid and host blood indicates that amino acids are accumulated to varying degrees in the cyst, the values obtained resembling those achieved during in vitro experiments. The relevance of these studies to drug targeting is discussed.

INTRODUCTION

Hydatidosis (echinococcosis, hydatid disease), caused by tapeworms of the genus Echinococcus, is a zoonotic disease of world-wide distribution which results in the development of hydatid cysts in the organs of humans and their domestic animals. Hydatid cysts do not generally respond to the commonly-available cestodicides and surgery is usually the treatment of choice. Although the surgical removal of cysts may be very successful for the large, well-developed unilocular cysts of Echinococcus granulosus (5), it is less successful in controlling the alveolar form of the disease due to E. multilocularis. In recent years, some chemotherapeutic success has been achieved with certain benzimidazoles, notably albendazole and mebendazole (see (1) for review). However, it must be concluded that the effects of these drugs are variable and insufficient to achieve cure in some patients and this failure seems to be primarily related to inadequate drug concentration being achieved in plasma and in parasite cysts (1). The search for a truly effective anti-cystic agent is thus far from over. In relation to chemotherapy, the question of permeability of the parasite is of paramount importance. At Keele, we are studying the transport of a variety of low molecular-weight nutrient molecules and macromolecules, not only in an attempt to fill this gap in our knowledge but also in the hope that parasite-specific transport systems may be revealed which could lend themselves to exploitation by more rational approaches to chemotherapy, such as drug-targeting. This paper summarises the results of studies concerning the transport of amino acids, glucose and a synthetic polymer (PVP) by Balb/c mouse-derived secondary hydatid cysts and horse hydatid protoscoleces.

METHODS & RESULTS

(a) Protoscolex

Protoscoleces were removed aseptically from horse liver hydatids, pre-incubated for 1h at 37°C in Hanks' saline then incubated for the appropriate time in a solution of ^{14}C-amino acid (plus inhibitors, where appropriate) in Hanks' saline at 37°C. Uptake was terminated by the addition of ice-cold saline and absorbed isotope assayed from alkaline digests of parasite material by liquid scintillation techniques.

Protoscoleces were found to absorb, by a combination of mediated transport and passive diffusion, 10 L-amino acids (cycloleucine, α -aminoisobutyric acid (AIB), alanine, leucine, proline, serine, phenylalanine, methionine, glutamic acid and lysine), D-alanine and D(+) glucose. All of these compounds were accumulated within the protoscolex, and the naturally occurring L-amino acids were incorporated into protein. L-alanine and L-methionine were metabolised into other (unknown) compounds. The kinetic parameters describing the uptake of these compounds are listed in Table 1. An interesting finding from the kinetic experiments is that, for neutral L-amino acids, a low number of transport loci (given by V_{max} values) for any particular compound appear to be compensated for by their high affinity (K_t values) for that compound. There does not appear to be any relationship between the magnitude of the diffusion component of transport and any kinetic or metabolic parameter. Inhibition studies reveal that the 10 L-amino acids enter the protoscolex through four, kinetically distinct, transport systems: one each for basic and acidic and two for neutral amino acids (one of the latter being highly specific for alanine, proline and to a lesser extent serine, and one for all except alanine). A separate study has shown that D-alanine is preferentially absorbed over its L-isomer. Not only is it transported at a faster rate and accumulated to a greater degree, but the transport system(s) which it uses are more numerous and have greater affinities (i.e. lower K_t's). This most unusual finding may have chemotherapeutic potential, in so far as the D-amino acid receptor, being parasite specific, may be used as a drug targeting site (Figure 1).

(b) Hydatid cysts.

Secondary hydatid cysts were obtained from Balb/c mice infected 300-400 days previously by intraperitoneal injection of c. 5000 protoscoleces removed aseptically from horse liver hydatid cysts. Previous experiments have shown that uptake is dependent on cyst size and is best expressed in terms of unit surface area. Cyst mass cysts, which are free of host response (3), of 3-8mm diameter ONLY were used. Following a 30 min pre-incubation at 37°C in Hanks' saline, cysts were incubated for the appropriate time in a solution of isotope, inhibitors where appropriate and Hanks' saline also at 37°C. After washing, a sample of hydatid fluid was removed from each cyst, and absorbed the isotope in the cyst wall extracted with ethanol. Assay of radioisotope was by liquid scintillation techniques.

Unlike the protoscolex, the hydatid cyst contains two major structural regions: the cyst wall (CW), consisting of a non-living laminated layer and the syncytial germinal layer, and a fluid-filled

Table 1.   Results of experiments involving protoscoleces.

Compound	Accumulation	KINETIC PARAMETERS K_t[b]	V_{max}[c]	K_d[d]	METABOLIC PARAMETERS in % Protein[e]	% Metabolised[f]
Clyco	5.7	0.124	0.947	0.240	(h)	(h)
AIB	ND	0.039	0.139	0.280	(h)	(h)
L-pro	4.0	0.009	0.055	0.185	7.3	0
L-ala	2.9	0.047	0.075	0.144	2.6[i]	24
L-leu	4.2	0.003	0.046	0.537	11.2	0
L-ser	3.0	0.024	0.089	0.424	20.5	0
L-phe	3.5	0.103	0.124	0.280	10.3	0
L-met	8.1	0.107	0.394	0.152	9.8	87
L-lys (g)	1.9	0.067	0.131	0.132	9.3	0
L-glu (g)	1.8	0.028	0.011	0.054	11.2	0
D-Ala	6.2	0.033	0.017	0.176	1.4(NS)[i]	21[i]
D(+)glu	4.5	0.214	0.041	0.050	(ND)	(ND)

a - after 30 min; b - units are mM; c - units are nmoles/mg protein/2 min;
d - units are nmoles/mg protein/2 min/mM; e - % of absorbed ^{14}C-amino acid incorporated into protein after 30 min incubation;
f - % of absorbed ^{14}C-amino acid metabolised after 30 min incubation;
g - at pH = 7; h - synthetic amino acids - not metabolised;
i - after 1h incubation;
ND - experiment not performed;   NS - not significant.

Fig. 1.   TARGETING TO THE D-AMINO ACID SITE ON E. GRANULOSUS

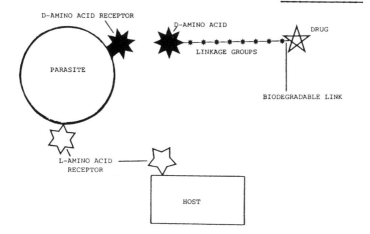

interior (HF). Studies involving incubation periods of up to 5h, showed that all of the amino acids studied, plus ^{14}C-L-threonine, were accumulated in HF and CW after 1 or 2h incubations. Further work with L-threonine, involving short term kinetic and inhibition studies of both the CW and HF components, showed that, despite the evidence of the time-course studies, entry into the CW component appears to occur solely by passive diffusion, and, in agreement with the time-course studies, entry into the HF component is by mediated transport with a K_t of 0.45mM and a V_{max} of 0.013 nmoles/mm^2/2 min. Thus it would appear that the results for the CW component are contradictory. A possible explanation is that the contribution made by the much larger laminated layer to solute flux is of such magnitude and probably passive in nature, that it swamps any mediated transport occurring in the germinal layer. An examination of this hypothesis awaits the development of a successful method for separating germinal and laminated layers post-incubation. Work is actively continuing in this area.

The hydatid cyst is also capable of absorbing PVP, albeit at a very slow rate. From time-course studies of uptake, an endocytic index of 0.004μl/mg protein/h can be calculated (c.f. value of 2μl/mg protein/h with rat yolk sac (6). The uptake of PVP was also shown to be temperature-dependent and inhibited by 2-,4-dinitrophenol, suggesting that the mechanism of uptake is an active, energy-requiring process.

(c) Amino acid analysis of host blood and hydatid fluid.

To place the results obtained with both the hydatid cysts and protoscoleces used during in vitro transport experiments into a physiological perspective, amino acid analyses of host blood plasma (horse and Balb/c mouse) and hydatid fluid obtained from cysts within the same host (horse and Balb/c mouse) are currently being performed. After de-proteinisation, samples of cyst fluid and plasma were analysed using a Jeol (Model JLC 6AH) automatic amino acid analyser. To date, we have sufficient data from the Balb/c mouse/mouse hydatid symbiosis to suggest that the majority of the 23 amino acids present are accumulated within the cyst in vivo, three are in equilibrium with plasma, and one (arginine) appears to be excluded (Figure 2). The TOTAL concentration of amino acids within the cyst fluid is 23.4mM and in blood 4.9mM, representing a concentration factor of 4.8x in the cyst fluid. Work is currently in progress to analyse the amino needs of the horse/horse hydatid symbiosis.

DISCUSSION

The most striking feature of this study is the marked difference between the abilities of hydatid cysts and protoscoleces to absorb certain nutrient molecules. For example, after 30 min incubation, none of the amino acids studied were accumulated by the hydatid cyst, but are accumulated by up to 5 x the medium concentration by the protoscolex, although similar levels of accumulation are achieved by both components of the hydatid cyst after 1 or 2h. A calculation of flux rates for cycloleucine, L-proline and L-alanine uptake by the hydatid cyst gives values of 5.1×10^{-3}, 2.2×10^{-3} and 18×10^{-3} μmoles/cm^2/h respectively, which are low in comparison with those of the protoscolex - with values of 1592×10^{-3}, 409×10^{-3} and 324×10^{-3} μmoles/cm^2/h respectively. These extremely slow rates of flux, coupled with poor short-term accumulative

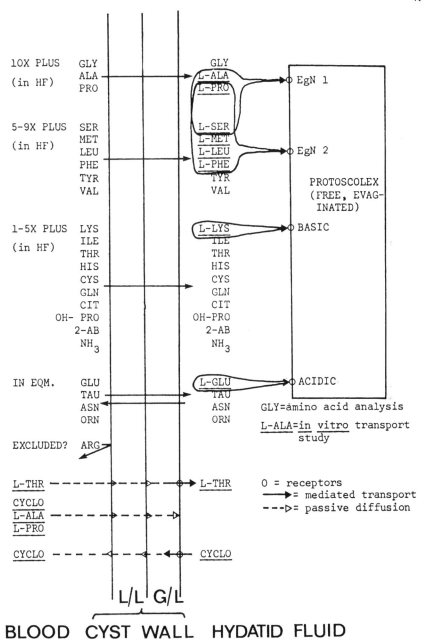

Fig 2. Amino acid fluxes in E.granulosus (horse strain). Blood and
        hydatid fluid are from Balb/c mice, protoscoleces from horse
        liver hydatids.

ability and the variability between cysts with regard to their ability to absorb the compounds under study (results not presented), may partially explain the unpredictable efficacy of the currently-available anti-cystic drugs. With regard to the protoscolex, it is important to recognise that under the experimental conditions employed, the protoscoleces evaginate. As the tegument of the evaginated worm differs from that of the non-evaginated organism (4) it is possible that we are, in effect, studying the uptake of compounds by the presumptive ADULT worm. Work is currently in progress to examine this hypothesis. The transport studies with L-amino acids do not indicate that the protoscolex receptors for these compounds are suitable targets for chemotherapy, although the unique presumptive D-amino acid system still holds promise. The PVP transport study indicates that it can be absorbed in vitro by hydatid cysts and it is proposed to investigate this further using other macrmolecules which are known to be present inside hydatid fluid, such as host immunoglobulins and serum albumin (see review by (2)). The preliminary results from the amino acid analyses of host blood and cyst fluid indicate that the concentration of amino acids achieved in vitro by the cyst, to some extent match those found in vivo, and that some, but by no means all, of the L-amino acid transport systems of the protoscolex are well-adapted to the levels of amino acids found within hydatid cyst fluid.

ACKNOWLEDGMENTS

We wish to thank the staff of Turner's Abbatoir, Nantwich, U.K. and Cooper's Abbatoir, Huddersfield, U.K. for the supply of horse hydatid cysts. The financial assistance of The Wellcome Trust, Smith Kline & French and the W.H.O. Parasitic Diseases Programme is gratefully acknowledged.

REFERENCES

1.) Eckert J. Prospects for the treatment of the metacestode stage of Echinococcus. In "The Biology of Echinococcus and Hydatid Disease". (ed. R.C.A. Thompson), pp. 250-284. George Allen & Unwin, London, 1986.

2.) McManus DP, & Bryant C: Biochemistry and physiology of Echinococcus. In "The Biology of Echinococcus and Hydatid Disease" (ed. R.C.A. Thompson), pp. 114-142. George Allen & Unwin, London, 1986.

3.) Richards KS, Arme C & Bridges JF: Echinococcus granulosus equinus: an ultrastructural study of murine tissue response to hydatid cysts. Parasitology, 86: 407-417, 1983.

4.) Richards KS & Rogan MT: Surface ultrastructure of a strobilate form of the horse strain of Echinococcus granulosus cultured in a monophasic medium. Annals of Tropical Medicine and Parasitology, 80: 267-8, 1986.

5.) Schantz PM, Van den Bossche H & Eckert J: Chemotherapy for larval echinococcosis in animals and humans : report of a workshop. Zeitschrift fur Parasitenkunde, 67: 5-26, 1982.

6.) Williams KE, Kidston EM, Beck F & Lloyd JB: Quantitative studies of pinocytosis. I. Kinetics of uptake of ^{125}[I]polyvinylpyrrolidone by rat yolk sac cultured in vitro. Journal of Cell Biology 64: 113-122, 1975.

DIAGNOSIS OF THE OVINE HYDATIDOSIS BY IMMUNOELECTROPHORESIS

F.MARTINEZ-GOMEZ, S.HERNANDEZ-RODRIGUEZ AND F.SERRANO-AGUILERA.

1. ABSTRACT

The sera from 130 ovine killed in the city slaughterhouse of Cordoba (Spain) were studied. The blood was obtained during the killing of the animals and the serum recovered by the routine procedures was maintained at -20°C in the laboratory until its use.

In the slaughterhouse these killed animals were studied in order to determine the presence of hydatid cysts, their localization, their fertility and viability, and the presence of some other parasitic or infectious diseases.

The antigen was obtained from ovine hepatic cysts which were fertile and viable. The hydatid fluid, which was obtained by puncture in aseptic conditions, was centrifuged and dialyzed at 4°C for 48 hours in distilled water. The final product was lyophilized and maintained under refrigeration until use.

The immunoelectrophoresis (IEP) was performed according to Varela et al. (6) in agarose plate and the results were statistically analyzed by the X^2 test.

Of the 130 animals studied, 67 were apparently healthy, 49 showed macroscopic hydatid cysts and the other 14 presented different infectious or parasitic conditions.

The specificity and sensitivity of the IEP test in the diagnosis of the ovine hydatidosis are based on the number of arcs present in each sample and not specifically related to the presence of the "arc 5".

We related our results to the fertility and viability of the hydatid cysts, their localization and the amount of hydatid fluid contained in each cyst. The presence or absence of some other parasitic conditions are also considered.

A maximum of specificity of 100% was found with the presence of 3 to 4 arcs, and a minimum of 76.64% with the presence of only 1 precipitation arc. The diagnostic value of the "arc 5" showed a sensitivity of 32.65% and a specificity of 98.76%.

2. INTRODUCTION

Hydatidosis is one of the most important zoonoses in Spain which has sanitary, economical and sociological repercussions. It is thus necessary to establish and evolve means to eradicate the disease. The attempts to control it have been considered difficult, among other factors, because of the limitations in the diagnosis of the disease in the intermediate host. The serological techniques used in living animals present variable results; on the one hand because of its low sensitivity mainly due to weak immunological response of the intermediate host (3) and on the other hand because of its low specificity due to the crossed reactions produced by other parasites (1). In the present study we have tried to establish the diagnostic criteria in sheep through immunoelectrophoresis (IEP and the factors influencing the interpretation of the results are also considered.

3. MATERIALS AND METHODS

We studied the sera from 130 sheep at the Municipal Slaughterhouse in Cordoba (Spain). The sera, obtained by the routine procedure, were stored at -20°C until used.

Also at the slaughterhouse we examined the killed animals to determine the presence or absence of parasitic infections especially those located in the peritoneum, liver or lungs.

The presence of hydatid cysts was determined by macroscopic examination and palpation of the different viscera; the number of the cysts, their localization, stage of development and volume were noted. The hydatid fluid was obtained by aseptic puncture, and we determined in our laboratory the fertility and viability of the cysts according to the technique described in a previous work (4).

The hydatid fluid antigen was obtained from fertile and viable liver cysts and was processed by centrifugation, dialysis and posterior lyophilization, and stored at 4°C until used. The control serum was obtained from a 6 month old sheep hyperimmunized with hydatid antigen according to Varela et al. (7).

The IEP was carried out in agarose plates (Filmuno-plates Operon) according to Varela et al. (6), and the results obtained were statistically analyzed by means of the $\underline{X}^2$ test.

4. RESULTS AND DISCUSSION

Of the 130 sera included in this study, 67 were from healthy animals, 49 (37.69%) were from animals with macroscopically detectable hydatid cysts (22 of these animals showed other parasitosis such as fascioliasis, dicrocoeliasis, dictyocaulosis or protostrongylosis and the other 14 were without hydatidosis but with other infectious or parasitic infections.

Table 1 shows the results obtained with the IEP test in relation to the number of arcs of precipitation. We have considered the diagnostic value of the "arc 5" and afterwards the diagnostic value of the number of other arcs present in the sample: four, three, two or only one arc. The results are referred to the 49 animals with hydatidosis and in the Table we show the percentage of sensitivity (number of positive cases/number of animals with Hydatidosis x 100) and the specificity (number of true negative cases/number of animals with hydatidosis x 100). This problem of the specificity of the "arc 5" has been studied by Rikard (5) in human hydatid disease.

With the $\underline{X}^2$ test we found a 99.95% of statistical significance of the relationship between sensitivity, specificity and criteria applied to the positivity, but we did not find significant differences when we consider the specificity obtained using the presence of the "arc 5" as diagnostic criterion or the presence of three other arcs. So the presence of three or more arcs should be considered as diagnostic value in the hydatid disease. Hamel & Ris (2) didn't find "arc 5" as a suitable test to the immunodiagnosis of ovine hydatidosis.

When we used the presence of a lesser number of precipitation arcs as criterion of positivity, we found a considerably higher sensitivity, but a lower specificity. In the case of two arcs, the sensitivity is 51.02% and the specificity 96.29%, so we could consider this result as valuable in the diagnostic of the hydatid disease.

TABLE 1

Results of the IEP test

N° of Arcs	+	F +	-	F -	Sen.%	Sp.%
"Arc 5"	17	1	113	33	33.65	98.76
4 or more	6	0	124	43	12.24	100
3	13	0	117	36	26.53	100
2	28	3	102	24	51.02	96.29
1	62	19	68	6	87.75	76.54

Note: +, positive; F+, False positive; -, Negative; F- False negative.

In Table 2 we relate our IEP results to the fertility and viability of the cysts. Fertility or infertility has no influence in the appearence of"arc 5", but we found more false negatives in the animals with infertile cysts, with an 80% of statistical significance. The animals with fertile cysts seem to present more than three arcs, with an 80% of statistical significance.

TABLE 2

IEP in relationship with the fertility
and viability of the cysts
(Expresed in number of cases and in %)

N° of arcs	Infertile		Fertile		Viable		Nonviable	
"Arc 5"	9	32.14	7	33.33	5	31.25	2	40.00
4 or more	2	7.14	4	19.04	4	25.00	0	
3	3	10.71	4	19.04	3	18.75	1	20.00
2	8	28.75	4	19.04	4	25.00	0	
1	10	35.71	8	38.09	4	25.00	4	80.00
0	5	17.85	1	4.76	1	6.25	0	

In Table 3 we show the relationship between the results obtained by the IEP and the evolutive stage of the cyst, according to the presence of hyaline hydatid fluid or the cases of cysts in processes of calcification or caseation.

We found a higher presence of "arc 5" in animals with calcified or degenerated cysts (14 from 16 examined animals) than in animals with hyaline cysts (2 from 16). The statistical level of significance is 80% according to the $\underline{X}^2$ test.

TABLE 3
(Results in number and in %)

N° Arcs	Hyaline		Cal./Cas.	
"Arc 5"	2	16.66	14	37.83
4 / more	1	8.33	5	13.51
3	1	8.33	6	16.21
2	2	16.66	10	27.02
1	5	41.66	13	35.13
0	3	25.00	3	8.10

The IEP results were compared in relation to the amount of hydatid fluid present in the cysts. In this way, we considered cysts with more than 10 ml of hydatid fluid and cysts with less than this volume. These results are shown in Table 4, and from these we can conclude that the volume of the cyst is related with the appearance of more arcs of precipitation, and higher appearence of the "arc 5" of Capron. Statistically these results present a 95% of significance.

TABLE 4
(Results in number and in %)

N° of Arcs	> 10 ml		< 10 ml	
"Arc 5"	8	44.44	8	25.80
4 or more	5	27.77	1	3.22
3	4	22.22	2	9.67
2	5	27.77	7	22.58
1	4	22.22	14	45.16
0		0	6	19.35

In the case of cysts with more than 10 ml of hydatid fluid we never found false negative reactions. That seems to demonstrate a direct relationship between the volume of the cyst and the level of host immunoligical response.

We studied in the same way the results of the IEP according to the localization of the cysts: hepatic, pulmonary or both. These results are presented in Table 5. We found the pulmonary localization seems to facilitate the appearance of the "arc 5", but the statistical significance of this data is only of 80%. In all the other cases there is no significant relationship between the presence of arcs and the localization of the cysts.

48

TABLE 5
IEP and localization of the cysts

(Expressed in number and in %)

N° of Arcs	Hepatic		Pulmonary		Mixed	
"Arc 5"	3	23.07	6	54.54	7	28.00
4 or more	3	23.07	0	0	3	12.00
3	0	0	2	18.18	5	20.00
2	3	23.07	4	36.36	5	20.00
1	4	30.37	5	45.45	9	36.00
0	3	23.07	0	0	3	12.00

In the Table 6 we have compared the results found in animals with hydatidosis but without other parasitosis with those obtained in animals harbouring hydatid cysts and other parasites like Fasciola, Dicrocoelium, Dictyocaulus, etc. The presence of these other parasites influences the results with a level of 99% significance.

TABLE 6
Influence of other parasite infections
(Expresed in number and in %)

N° of Arcs	Parasitized		Non. Par.	
"Arc 5"	5	22.72	11	40.74
4 or more	2	9.09	4	14.81
3	2	9.09	5	18.51
2	10	45.45	2	7.40
1	5	22.72	13	48.14

The concurrence with other parasites determines lower appearence of the "arc 5" and of the three or more arcs. The higher positive results in animals with polyparasitism were found using the presence of two arcs as a criterion of positivity, which could be interpreted as the existence in these cases of an additional antigenic stimulus producing cross reactions.

We only found a false negative reaction in an animal with fertile and viable cysts, but the volume of the cysts was very small. All the other false negative cases were found in animals with infertile cysts.

AKNOWLEDGEMENT. This work was supported by the CAICYT through grant PR84-0270-C03-01.

REFERENCES

1. Conder GA, Andersen FL and Schantz PM: Immunodiagnostic tests for hydatidosis in sheep: an evaluation of double diffusion, immunoelectrophoresis, indirect haemagglutination, and intradermal tests. Journal of Parasitology, 66: 577-584, 1980.

2. Hamel KL and Ris DR: The use of a cathodic antigen in the immunoelectrophoretic diagnostic of Echinococcus granulosus in sheep. Veterinary Immunology and Immunopathology, 3: 419-425, 1982.

3. Judson DG, Dixon JB, Clarkson MJ and Pritchard J: Ovine hydatidosis: some immunological characteristics of the seronegative host. Parasitology, 91: 349-357, 1985.

4. Martinez-Gomez F, Henandez-Rodriguez S, Navarrete I and Calero-Carretero R: Serological tests in relation to the viability, fertility and localization of the hydatid cysts in cattle, sheep, goats and swine. Veterinary Parasitology, 7: 33-38, 1980.

5. Rickard MD: Serological diagnosis and post-operative surveillance of human hydatid disease. I. Latex agglutination and immunoelectrophoresis using crude cyst fluid antigen. Pathology, 16: 207-210, 1984.

6. Varela-Diaz VM, Coltorti EA, Ricardes MI, Guisantes JA and Yarzabal LA: The immunoelectrophoretic characterization of sheep hydatid fluid antigens. American Journal of Tropical Medicine and Hygiene 23, 1092-1096, 1974.

7. Varela-Diaz VM, Guisantes JA, Ricardes MI, Yarzabal LA and Coltorrti EA: Evaluation of whole and purified hydatid fluid antigens in the diagnosis of human hydatidosis by the immunoelectrophoresis test. American Journal of Tropical Medicine and Hygiene, 24: 293-303, 1975.

CHARACTERISATION OF *ECHINOCOCCUS GRANULOSUS* PROTEINS AND ANTIGENS FROM
HYDATID CYST FLUID.

J.C.SHEPHERD , D.P.MCMANUS

ABSTRACT
    To date, the antigens of the cyst fluid of *Echinococcus granulosus*
have been little studied, apart from partial isolation and characterisation of the major antigens, antigen 5 and antigen B. As a prerequisite to
choosing a suitable candidate antigen for detection in the circulation of
infected individuals, we have characterised cyst fluid proteins with
respect to relative molecular mass and antigenicity using immunoprecipitation analysis, SDS-PAGE and immunoblotting. A PAS staining molecule of
38K, which migrates with a different M_r under non-reducing conditions, has
been presumptively identified as the large sub-unit of Antigen 5. Moreover,
bands at 20, 16 and 12K probably represent the sub-units of Antigen B.
Other antigens are also present in cyst fluid but these cannot be identified easily and consistently in different samples of fluid and all molecules present in cyst kluid in high concentration are immunogenic in
humans. Antigen 5 and Antigen B represent the most concentrated parasite-
derived proteins in cyst fluid and so for serodetection one or other of
these antigens will probably have to be used. Antigen B is known to be
less immunoreactive than Antigen 5 and thus if an *E. granulosus*-specific
epitope can be selected from Antigen B, it may well prove to be an appropriate target for detection purposes.

1. INTRODUCTION
    Unilocular hydatid disease, caused by the dog tapeworm, *Echinococcus
granulosus*, is recognised as one of the world's major zoonoses, affecting
both man and his domestic animals. The disease in humans is caused by
infection with the onchosphere which subsequently develops into a fluid
filled cyst containing larval worms, or protoscoleces. The pathology
associated with the disease is mainly due to the physical pressure the
developing cyst exerts on the host's viscera (primarily the liver or the
lungs). Anaphylaxis may also occur when a cyst ruptures and releases large
volumes of cyst fluid containing parasite antigen into the sensitised host.
There is rarely any parasitological evidence of infection and clinical
symptoms are non-specific. Consequently, serodiagnosis of the disease has
received much attention.
    The present most widespread method of using sheep cyst fluid as an anti-
gen to detect circulating antibody by various tests such as Immunohaema-
glutination, Immunodiffusion, Immunoelectrophoresis and Enzyme Linked
Immunosorbent Assay has been of clinical and seroepidemiological use (1).
However these tests suffer from lack of sensitivity due to a low antibody
response in some individuals and immunocomplexing of antibody available
for detection (2, 3). In addition, they lack specificity due to the cross-
reactivity of *E. granulosus* antigens with those of other helminths (3-5).

The cyst fluid in which the protoscoleces develop is rich in soluble proteins, some of which are derived from the host and some are excreted or secreted by the parasite. The cyst wall, comprising several layers of both host and parasite origin, is permeable to host serum proteins, the most notable of which are serum albumin and IgG. These serum proteins can be detected in cyst fluid (6, 7). The cyst wall also allows the passage of parasite antigens out of the cyst into the circulation and these have been detected using a rabbit antiserum to 'capture' antigen in a two site assay, thereby improving the sensitivity of serodiagnosis over antibody detection tests and reducing the number of false negative results obtained (2). Circulating antigen has also recently been detected by western blotting of dissociated human immune complexes (3). Thus it is possible to detect circulating antigen in hydatid disease using a polyclonal reagent but it may be possible to improve the signal to noise ratio of the test further by examining cyst fluid, the pool of circulating antigen, and selecting an *E. granulosus* specific antigen that is always produced by the parasite but is not immunocomplexed. However the antigens of cyst fluid have been little studied, apart from partial isolation and characterisation of the major antigens, antigen 5 and antigen B (notation of Rickard and Lightowlers, see Reference 8). Thus, as a prerequisite to choosing an antigen suitable for detection, we have characterised proteins of cyst fluid by relative molecular mass (M_r) and antigenicity by using immuno-precipitation and sodium-dodecyl-sulphate polyacrylamide gel electro-phoresis (SDS-PAGE) and immunoblotting.

2. RESULTS

Figure 1 shows the iodine/iodogen labelled antigens in sheep cyst fluid recognised by a pooled human infection serum taken from individuals in Kenya and the U.K., compared with total and TCA precipitated labelled proteins.

Macromolecules can be discerned under reducing conditions with M_r 12, 16, 20, 25, 38, 40, 49, 56, 66, 85 and 125 KiloDaltons (K); all were pre-cipitable by 10% trichloroacetic acid, indicating that they contain an element of protein. There may be many more macromolecules that are not labelled by the iodogen technique or cannot be resolved in one dimension. The Coomassie Blue stained lane indicates that the proteins present in the highest concentration are the 12, 16, 20, 38 and 68K molecules although there are numerous other minor bands. In addition, the 16, 38, and 68K bands stained strongly with Periodic Acid/Schiff's Reagent (PAS) (data not shown) indicating that they are glycosylated.

Immunoprecipitation of labelled antigens with the human antiserum pool, and separation under reducing conditions, revealed antigens with M_r 12, 16, 20, 24, 28, 38, 43, 66, 68, 76, 85, 165, 180, 200 and 230K. All the major proteins present in cyst fluid observed in this study were precipitated by human antibodies to *E. granulosus*. The increased number of bands obser-ved following immunoprecipitation was possibly due to selection and con-centration of low abundance antigens. In the absence of 2-Mercaptoethanol (2-ME), which reduces disulphide bonds to sulphydryl groups, most of the antigens migrated with the same apparent M_r with the exceptions of the 38, 66, 68, 76 and 85K molecules. These molecules are therefore likely to be subunits of larger components bound together by disulphide bonds. A heavily iodinated molecule occurs at 61K which is presumably host serum albumin, a downward shift in M_r being characteristic of this molecule when not reduced.

52

Additional antigens were apparent at M_r 100, 110, 135, 170, 230 and 270K, together with bands in the stacking gel and sample well which are either of very high M_r or insoluble in the absence of 2-ME. Immunoprecipitation with normal human serum revealed bands at 20, 38, 50 and 68K under reducing conditions, although the 20 and 38K bands were very faint. The 50 and 68K bands are probably IgG heavy chain and serum albumin, normal human serum containing antibodies to ingested sheep meat proteins (R.M.Maizels, pers. comm.). Antigen B and antigen 5 are probably the bands at 38, 20, 16 and 12K, the 38K band being the large subunit of antigen 5 and the small subunit being occluded by the subunits of antigen B which are heavily labelled with iodine. Under non-reducing conditions, the two subunits of antigen 5 associate by disulphide bonding and the resultant single subunit is probably masked by the heavily labelled group of proteins at 60-70K.

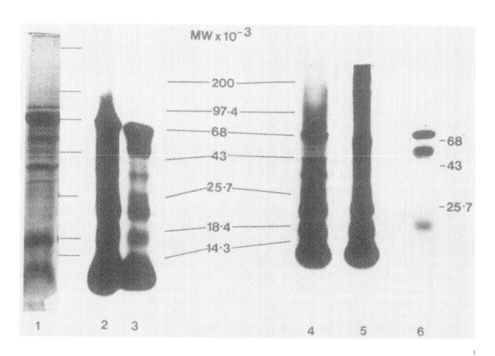

FIGURE 1. Macromolecules and antigenic macromolecules in cyst fluid from sheep. Lanes 2-6 show proteins and antigens labelled by the Iodogen method (20) and separated on 5-20% polyacrylamide gradient slab gels using the discontinuous buffer system of Laemmli (21). Immunoprecipitations utilised formalin-fixed *Staphylococcus aureus* (22). Fixed gels were autoradiographed at -70° C. Proteins in lanes 1-4 and 6 were separated under reducing conditions.Proteins in lane 5 were not reduced.
Lane 1 - Cyst fluid (100ug protein) separated as for lanes 2-6 and stained with Coomassie Blue. Lane 2 - Total labelled proteins . Lane 3 - Trichloroacetic acid precipitated labelled proteins. Lane 4 - Labelled antigens precipitated with pooled human infection serum. Lane 5 - Same as 4.
Lane 6 - Labelled antigens precipitated with pooled normal human serum.

The weak precipitation of antigen 5 by normal human serum may be arte--
factual as it was not bound by normal human serum on an immunoblot (see
Figure 2). The most heavily stained antigen on a blot (Figure 2) is the
38K subunit of antigen 5, human infection serum presumably containing a
high concentration of antibody and/or antibody of high affinity to this
antigen. Under non-reducing conditions there was a heavily stained region
of M_r greater than 200K which was not resolved into bands. In addition,
antigen B was bound, as were a number of antigens with M_r of 80K or more
under reducing conditions.

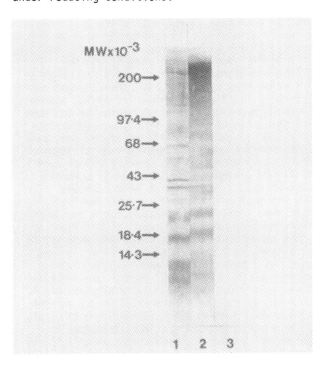

FIGURE 2. Antigens of sheep cyst fluid separated as described for Figure 1
and then transferred to nitrocellulose paper (23). Nitrocellulose strips
were quenched, incubated in 1/100 pooled human serum and then incubated
with peroxidase conjugated second antibody and visualised with diamino-
benzidine substrate. Strip 1 - Reduced proteins probed with pooled human
infection serum. Strip 2 - Unreduced proteins probed with pooled human
infection serum. Strip 3 - Reduced proteins probed with pooled normal
human serum.

3. DISCUSSION

The two major antigens of hydatid cyst fluid, antigen B and antigen 5,
are bound by a human antiserum pooled from U.K. and Kenya cases and they
can be immunoprecipitated and immunoblotted from every sample of sheep
hydatid fluid that we have tested so far (data not shown). Antigen 5 is
the antigen detected in the Arc 5 immunodiffusion test and was thought to
be specific to *E. granulosus* (9) until several workers reported anti-
antigen 5 antibodies in humans and animals infected with other cestodes

(10,11). It is an iodinatable, Concanavalin A binding lipoprotein of
varying high molecular weight and comprising of subunits of 60-70K which
themselves dissociate on reduction of disulphide bonds to two subunits of
circa 20K and circa 40K (12-14). In this study, a PAS staining subunit of
38K which migrates with a different M_r when not reduced has been identified
presumptively as the large subunit of antigen 5.

Antigen B has received less attention than antigen 5 as antibodies to
it are less easy to detect in infection serum (13) and it is thought to be
non-specific (9). It is an iodinatable, Concanavalin A binding lipoprotein
of approximately 150K, comprising three subunits of approximately 12K, 16K
and 20K, which are unaffected by reduction of disulphide bonds (12-14).
These M_r values are in accordance with those identified for antigen B in
this study. In addition another antigen with a serum albumin like mobility
has been identified in cyst fluid by immunoelectrophoresis but not
characterised further (15). Other antigens are also present, as shown in
this preliminary study and from work by Craig (3), but these cannot be
identified easily or consistently in different samples of fluid, possibly
due to their susceptibility to degradation. No obvious molecule has been
detected by the methods used here that is present in high concentration
but not immunogenic in humans; such a molecule would be a good candidate
for detection in a circulating antigen assay because of lack of immunocomplexing. It would be difficult to establish the specificity of such
a molecule without parasite material from a wide spectrum of parasite
diseases to test for cross-reactivity. The Coomassie Blue staining
presented here suggests that the most concentrated parasite-derived
proteins in cyst fluid are antigen 5 and antigen B and so for serodetection
either of these antigens will probably have to be used. In the light of
previously published work, the specificity of both these antigens may be
insufficient to prevent the occurence of false positives (9-11) and so
antisera may have to be raised to portions of the molecule or a series
of monoclonal antibodies raised which recognise *E. granulosus*-specific
epitopes. Recently, monoclonal antibodies have been raised to antigen 5
(16) which on immunoblots give a non-reduced subunit size of 56 and 66K
(two bands) but will not bind to the reduced subunits on nitrocellulose,
suggesting that they recognise a conformational epitope destroyed by
separation of the disulphide bonded subunits. However the specificity of
these antibodies has not been tested.

Any *E. granulosus*-specific epitope selected from antigen 5 or antigen B
as a candidate for circulating antigen detection should be one that is not
saturated with antibody. Musiani *et al* (17) found that antigen B was
present in cyst fluid at ten times the concentration of antigen 5 but that
it was less immunoreactive with human antibodies. Thus, if a specific
epitope is identified, antigen B may be an appropriate target for detection.
Neither antigen 5 nor antigen B were apparent in the serum blotted by
Craig (3). Work in this laboratory indicates that these antigens are always
present in cyst fluid, regardless of source. Nevertheless, it is important
to ascertain whether every hydatid cyst in whatever internal location or
parasitological status and from every endemic area, contains these two
antigens and whether they are always able to leave the intact cyst and
enter the circulation. It has been found that there is more free circulating antigen from liver cysts compared with lung cysts but that immune
complexes are raised in lung cysts (2). Cyst fluid antigens recognised by
infection serum have been found not to change with different sources of
human antiserum (18). This suggests that the well documented strain

variation in *E. granulosus* (19) may not have to be taken into account when choosing a target antigen for detection.

The above work was performed using human sera as work in this laboratory indicates that it is very difficult to perform the same experiments using sheep and horse sera obtained from animals naturally infected with *E. granulosus*. "Normal" sheep and horses invariably harbour a variety of helminths possessing antigens that extensively cross react with *E. granulosus* cyst fluid antigens, and sera from animals infected with only one species of helminth are unavailable to us. However, it is possible that a specific *E. granulosus* antigen or epitope identified using human antisera will not cross-react with antigens of other animal parasites.

4. ACKNOWLEDGEMENTS

D.P.M. gratefully acknowledges financial support from Wellcome Trust, E.E.C. and M.R.C. for this work.

5. REFERENCES

1. Williams JF: Cestode Infections: In "Immunology of Parasitic Infection" eds. Cohen S and Warren KS, Blackwell Scientific Press, Oxford, pp.676-714, 1982
2. Craig PS and Nelson GS: The detection of circulating antigen in human hydatid disease. Annals of Tropical Medicine and Parasitology, 78: 219-227, 1984
3. Craig PS: Detection of specific circulating antigen, immune complexes and antibodies in human hydatidosis from Turkana (Kenya) and Great Britain. Parasite Immunology, 8: 171-188, 1986
4. Schantz PM, Shanks D and Wilson M: Serologic cross-reactions with sera from patients with echinococcosis and cysticercosis. American Journal of Tropical Medicine and Hygiene, 29: 609-612, 1980
5. Speiser F: Application of the enzyme-linked immunosorbent assay (ELISA) for the diagnosis of filariasis and echinococcosis. Tropenmedizin und Parasitologie, 31: 459-466, 1980
6. Chordi A and Kagan IG: Identification and characterization of antigenic components of sheep hydatid fluid by immunoelectrophoresis. Journal of Parasitology, 51: 63-71, 1965
7. Coltorti EA and Varela-Diaz VM: IgG levels and host specificity in hydatid cyst fluid. Journal of Parasitology, 58: 753-756, 1972
8. Rickard MD and Lightowlers MW: Immunodiagnosis of Hydatid Disease. In "The Biology of *Echinococcus*", Ed. Thompson RCA, George Allen and Unwin, London: 225-249, 1986
9. Bout D, Fruit J and Capron A: Purification d'un antigène specifique de liquide hydatique. Annales Immunologie Institut Pasteur: 125C, 775-788, 1974
10. Varela-Diaz VMJ, Eckert J, Rausch RL, Coltorti EA and Hess U: Detection of *Echinococcus granulosus* diagnostic arc 5 in sera from patients with surgically-confirmed *E. multilocularis* infection. Zeitschrift für Parasitenkunde, 53: 183-188, 1977
11. Yong WK and Heath DD: "Arc 5" antibodies in sera of sheep infected with *Echinococcus granulosus, Taenia hydatigena* and *Taenia ovis*. Parasite Immunology, 1: 27-38, 1979
12. Oriol R, Williams JF, Perez-Esandi MV and Oriol C: Purification of lipoprotein antigens of *Echinococcus granulosus* from sheep hydatid fluid. American Journal of Tropical Medicine and Hygiene, 20: 569-574, 1971

13. Pozzuoli R, Piantelli M, Perucci C, Arru E and Musiani P: Isolation of the most immunoreactive antigens of *Echinococcus granulosus* from sheep hydatid fluid. Journal of Immunology, 115: 1459-1463, 1975
14. Piantelli M, Pozzuoli R, Arru E and Musiani P: *Echinococcus granulosus*: identification of subunits of the major antigens. Journal of Immunology, 119: 1382-1386, 1977
15. Pezzella M, Galli C, Delia S, Vullo V, Zennaro F, Lillini E and Sorice F: Fractionation and characterization of hydatid fluid antigens with identification of an antigen similar to human serum albumin. Transactions of the Royal Society of Tropical Medicine and Hygiene, 78: 821-826, 1984
16. Di Felice G, Pini C, Afferni C and Vicari G: Purification and partial characterization of the major antigen of *Echinococcus granulosus* (antigen 5) with monoclonal antibodies. Molecular and Biochemical Parasitology, 20: 133-142, 1986
17. Musiani P, Piantelli M, Lauriola L, Arru E and Pozzuoli R: *Echinococcus granulosus*: specific quantification of the two most immunoreactive antigens in hydatid fluids. Journal of Clinical Pathology, 31: 475-478, 1978
18. Craig PS, Zeyhle E and Romig T: Hydatid disease: research and control in Turkana II. The role of immunological techniques for the diagnosis of hydatid disease. Transactions of the Royal Society of Tropical Medicine and Hygiene, 80: 183-192, 1986
19. McManus DP and Smyth JD: Hydatidosis changing concepts in epidemiology and speciation. Parasitology Today, 2: 163-168, 1986
20. Markwell MAK and Fox CF: Surface-specific iodination of membrane proteins of viruses and eukaryotic cells using 1, 3, 4, 6-tetrachloro-3α, 6 α - diphenylglycouril. Biochemistry, 17: 4807-4817, 1978
21. Leammli UK: Cleavage of structural proteins during the assembly of the head of bacteriophage T4. Nature, 227: 680-685, 1970
22. Kessler S: Rapid isolation of antigens from cells with a staphylococcal protein A-antibody absorbent: parameters of the interaction of antibody-antigen complexes with protein A. Journal of Immunology, 115: 1617-1624, 1975
23. Towbin H, Staehelin T and Gordon J: Electrophoretic transfer of proteins from polyacrylamide gels to nitrocellulose sheets: procedure and some applications. Proceedings of the National Academy of Sciences, USA, 76: 4350-4354, 1979

EXISTENCE OF AN URBAN CYCLE OF ECHINOCOCCUS GRANULOSUS IN
CENTRAL TUNISIA

A. BCHIR, A. JAIEM, M. JEMMALI, J.J. ROUSSET,
C. GAUDEBOUT, B. LAROUZE

ABSTRACT
     The prevalence rate of infection with Echinococcus
granulosus was measured in Central Tunisia in 50 dogs investigated by autopsy. Eleven dogs (22.0%) were infected: 2 of the
10 dogs from a rural area, 7 of the 10 dogs from an urban area
surrounding a garbage disposal pit, and 2 of the other 30 dogs
of urban origin. The present results suggest the existence in
Central Tunisia of an urban cycle of E. granulosus in addition
to the rural cycle.

INTRODUCTION
     Hydatidosis is a major public health problem in Central
Tunisia: the mean annual surgical incidence rate for 1984 was
22.8 per 100 000 inhabitants, varying from 5.0 to 52.1 per
100 000 according to the district (1). Recent serologic and
echotomographic screening in a high risk district showed a
prevalence rate of 3.6 per 1000 (5).
     In North Africa, the cycle of E. granulosus is considered to
be essentially rural involving predominantly the sheep as
intermediary host and the dog as definitive host (4). In
Tunisia, the population of dogs has been evaluated as 1 million
(one dog per 6 inhabitants)(2). In both urban and rural settings, there are numerous stray dogs which feed on garbage
littered by the population. Heavy concentrations of stray dogs
are observed in and around garbage disposal areas. In some of
these areas viscera which are unfit for human consumption are
discarded by the local slaughter house and only sprayed with an
antiseptic liquid. In urban settings, the domestic dogs are
usually kept indoors. In rural settings, domestic dogs are
released at night to deter stray dogs and wild carnivors (5).
Dogs take this opportunity to look for their meat ration
(discarded meat, dead animals which are often left without
being buried).
     In Tunisia, there are only a few reports on the prevalence
rate of Taenidae in dogs through standard faeces examination,
and no specific report on the prevalence rate of E. granulosus.

MATERIALS AND METHODS
     The prevalence rate of E. granulosus was studied in Central
Tunisia in 50 dogs which were shot at night. Ten dogs were

from a rural district at high risk located in the northern part of the Governorat of Sousse. The limited number of dogs in this subgroup was due to refusal by the local population since the investigated dogs included domestic dogs which were let out at night. The other 40 dogs were from a poorer district on the edge of the city of Sousse, the second largest city in Tunisia (99 500 inhabitants). In this district, 10 of the dogs were from an area located within 1 km of a garbage disposal pit. The other 30 dogs were obtained from the same district at more than 1 km from this disposal pit. The dogs were autopsied within 2 hours of killing. After ligature at both ends, the small intestine was removed and placed in 5% formalin water. The intestinal content was then scraped off. E. granulosus was identified under a binocular microscope (3). Statistical analysis was carried out using the χ^2 test with Yates correction for continuity.

RESULTS

Of the 50 dogs investigated, 11 (22.0%) were infected with E. granulosus: 2 of the 10 dogs from the rural area, 7 of the 10 dogs from the urban area surrounding the disposal pit, and 2 of the other 30 dogs of urban origin. In the urban area, the infection was more frequent in dogs killed around the disposal mit (χ^2_C = 15.9, df = 1, p < 0.001).

DISCUSSION

The present results strongly suggest that in addition to the rural cycle of E. granulosus in Central Tunisia, there is an urban cycle. The discussion of the figures observed should take into account the small size of the sample studied. Notwithstanding, the results on dogs killed around the disposal pit are consistent with the fact that there are areas at particularly high risk for hydatidosis within the large city of Sousse. Areas around this disposal pit, as is common in cities in North Africa, are crowded and inhabited by poor populations. The cluster of stray dogs due to the availability of discarded food, the high prevalence rate of dog infection due to access to infected viscera, and the poor hygienic conditions under which the populations live might amplify the risk of human infection. Despite the difficulty of identifying precisely the place of occurrence of a given human infection, the study of surgical incidence rates of urban hydatidosis might be instructive by possibly showing a cluster a cases in such a high risk area.

Measures for preventing hydatidosis are well identified and have proven effective particularly in Mediterranean countries such as Cyprus (6). Our figures show that much remains to be done in Tunisia, even in urban settings. As demonstrated by the difficulties encountered during our limited investigation in a rural setting, any measure involving the control of the dog population, would imply the participation of the community in addition to the legal measures.

REFERENCES

1. Bchir A, Jemni L, Allegue M, Hamdi A, Khlifa K, Letaieff R, Mlika N, Dridi H, Larouzé B, Rousset JJ, Gaudebout C and Jemmali M: Epidémiologie de l'hydatidose dans le Sahel et le Centre tunisiens. Bulletin de la Société de Pathologie Exotique, 78: 687-690, 1985.
2. Ben Osman F: La population canine en Tunisie. Séminaire sur le contrôle de la population canine en Tunisie. Ecole véterinaire de Sidi Thabet, Tunisie, 1981.
3. Euzéby J: Les échinococcoses animales et leurs relations avec les échinococcoses de l'homme. Paris: Vigot, 1971.
4. Larouzé B, Jemmali M, Yang R, Dihda L and Rousset JJ: Epidémiologie et prévention de l'hydatidose en Afrique. In Actualités en Pathologie infectieuse. Paris: Arnette, pp 191-214, 1981.
5. Mlika N, Larouzé B, Gaudebout C, Braham B, Allegue M, Dazza MC, Dridi M, Gharbi S, Gaumer B, Bchir A, Rousset JJ, Delattre M and Jemmali M: Echotomographic and serologic screening for hydatidosis in a Tunisian village. American Journal of Tropical Medicine and Hygiene, 35: 815-817, 1986.
6. WHO/FAO/UNEP: Guidelines for surveillance, prevention and control of echinococcosis hydatidosis. WHO/VPR/81.28, Geneva, 1981.

CONTROL OF ECHINOCOCCOSIS IN CYPRUS

K. POLYDOROU

BACKGROUND OF THE PROBLEM

During the dawning years of this century animal movements from and into the island were frequent and largely uncontrolled. Such interchanges of animals usually involved close neighbouring countries of Cyprus. These did not only include food animals but also dogs.

How, when and from where echinococcosis came to Cyprus is a matter of speculation. Once, however, the disease came to Cyprus its spread was greatly facilitated by the ways the Cypriots were associated with their animals and in general, the prevailing conditions on the island. Most of the people were farmers and each family kept its own small herd of sheep and goats and a few draught animals (native cattle, mules etc.) for working their fields and for transport. The sheep were kept in the yard of the house and the large animals often indoors during the long winter nights.

Viscera infected with
E.granulosus hydatid cysts

Of course this was also a way of keeping a close watch on the animals, since theft was common. To this effect large numbers of dogs were also kept in order to guard the owner's property including animal stock.

Slaughter of animals was done in the home-yards where the animals were kept and the viscera given as feed to the dogs.When animals were slaughtered at the village square the attendant dogs were also given unwanted pieces of offal.

The country-side was regularly sprinkled with carcasses of sheep and goats dying from diseases and the carcasses were largely left unburied and provided a natural food supply for a thriving population of stray dogs.

In the towns feeding the dogs was more of a problem. Usually table left-overs were given, if anything was given at all. In some instances, cheaper forms of offal, like "white liver" or lungs were given. Dogs in towns were seen every-where, but more frequently in the vicinity of abattoirs or refuse disposal areas.

Often they roamed in the fields around the towns in search of cadavers of food animals. thus, infection among these dogs was common and this was a definite hazard to the townspeople, who were, as a rule, unaware of the danger.

The Cypriots often went to considerable trouble to abandon the numerous unwanted puppies in some place where they stood a chance of survival. They were kindly disposed towards these animals and disliked killing them. Thus a serious stray dog problem was perpetuated.

At various times attempts were made to involve the police and the staff of other departments in controlling the number of stray dogs in rural areas by shooting; but the work was unpopular and outside their normal duties, and the natural level of the stray dog population was soon restored.

From time to time also committees were formed, propaganda campaigns were launched and anthelmintics were issued free of charge to dog owners. Generally, the impact of all these measures was negligible. The public apathetic; they loved their dogs, were sympathetic toward the strays and just did not want to know about hydatid disease. Among official bodies nobody felt it was their specific responsibility to deal with the problem, and the half-hearted and ineffectual attempts which were made did little to alter the situation.

At last in 1970, the whole problem was properly tackled for the first time. A 15-years plan for eradication was drawn up by the director of the Department of Veterinary Services approved ty the Government.

THE CAMPAIGN PLAN

In the plan all actions to be taken were outlined, arranged even in a timetable, whereby it was envisaged that in 15 years, echinococcosis would be completely eradicated from the island. Characteristic of the plan, was that the campaign was to be totally undertaken by the Veterinary Department, including all measures to be implemented in the campaign e.g. control and testing of dogs, control of slaughter, education of the public, assessment of the progress of the campaign and any other action which was deemed necessary for the success of the campaign. When required the Veterinary Department co-operates with other governmental services.

The methods used for the complete disruption of the cycle of echino-coccosis were based upon the local conditions prevailing in the island. Within a relatively short period it became possible to destroy all stray and ownerless dogs. Also, other measures were implemented so that the remaining dogs would come under control. Such measures included: (a) the obligatory registration of dogs with the authorities of the village or town as well as with the teams of the campaign against echinococcosis; (b) the spaying of as large a number of female dogs as possible so that dog repro-duction would be minimized; (c) the imposition of high registration fees for the ownership of unspayed female dogs, and (d) the obligatory testing on a continuous basis of all dogs for echinococcosis every three months using arecoline hydrobromide solution   for the detection of the E.granulosus taenia in the faeces of the dogs.

62

Mobile Testing Unit: Drenching dog with
Arecoline Solution

Regarding the animals/intermediate hosts of <u>Echinococcus</u> it was arranged
that their slaughter would be carried out always in approved slaughterhouses having an incinerator or pit where infected offal from slaughtered
animals are either burnt or thrown in. The entire compound is fenced so
that no dogs can enter the place.

One of the 210 approved Rural Slaughter-Houses

Regarding the measures implemented by the campaign and their objective
these were explained to the public before their implementation as well as
subsequently so that the public might positively contribute to the efforts
of the various teams of the campaign. Practical and easy to understand
methods were used in order to stir the interest of the public, the contribution of which was indeed invaluable for the success of the campaign.

RESULTS AND DISCUSSION

1. Dog Control

(a) Dog population

By the end of 1985 dog population was about 20,000 of these 12,000
were male and 8,000 bitches, compared to 50,000 before the campaign was
started. About 30% of the female dogs were spayed. The dog population
after 1977 started to increase as a result of the success of the campaign
and the feeling by the people that these animals if properly kept no longer
posed a danger for echinococcosis infection. Most dogs kept at present are
purebreds.

(b) Stray dog elimination

This was the measure given top priority in the campaign. From 1971-
1985, when the scheme was started about 86,000 stray dogs were destroyed
or on the average, about 6,000 per year.

(c) Spaying of bitches

From 1971 to 1985 about 13,500 bitches were spayed or on the average
about 900 per year. Unspayed bitches pay an annual registration fee of
$ 16 U.S. compared to $ 2 for spayed bitches and male dogs.

Figure I shows population changes and number of dogs exterminated from
1971 - 1985.

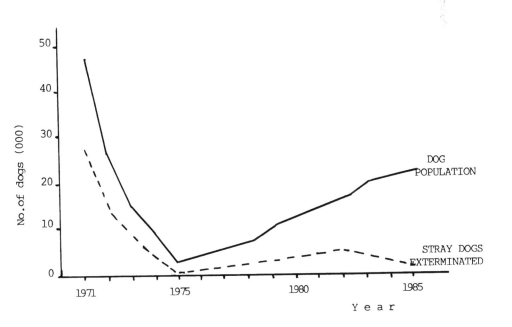

(d) Dog Testing

During the first examination of 1972 the infection level was 6.82%. By 1977 it dropped to 0.754% and by 1984 to 0.00% i.e. there was total elimination of infection.

Figure II shows infection in dogs throughout the Campaign.

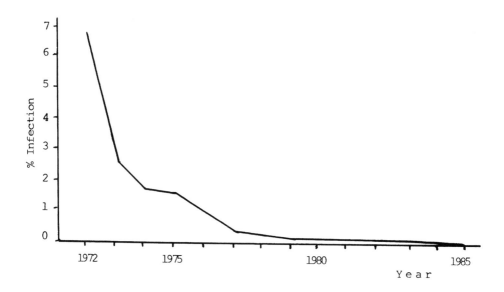

2. Infection in Farm Animals

Infection levels prior to 1971 were recorded from carcass inspection done at the municipal abattoirs at various intervals in the past. By 1976 infection in young animals was non-existent. This is an evidence of the fact that the environment was cleared from Echinococcus ova. In sheep over two years of age infection dropped from 66% prior to 1971 (when the Campaign started) to 0.11% in 1985 i.e. there was a reduction of 99.8%. From special surveys it was seen that the infected sheep in this age group were aged over ten years i.e.   animals born prior or during the first stages of the implementation of the Campaign. The same can be stated for mature goats and cattle. In goats infection dropped from 13.0% in 1971 to 0.00% in 1984 i.e. total elimination. In cattle respective numbers were 55.0% and 0.00%.

Figure III shows infection in mature sheep, goats and cattle from 1971-85.

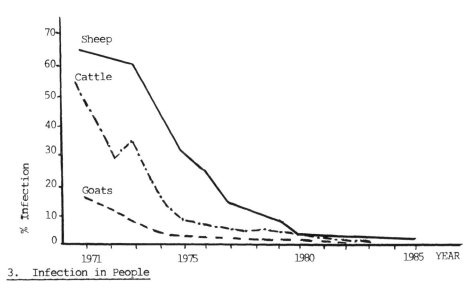

3. Infection in People

The disease was very common among the Cypriots of all ages. There was no village without a known victim, and in some families two to three members were affected. Instances of multiple surgery were common. For decades before the Campaign there were 50-60 operations/year among a population of 500,000. The annual surgical rate from 1961-70 was about 11/100,000. In 1979, 15 cases were reported by the Ministry of Health, giving an incidence of about 3/100,000 and in 1980 8 surgical cases 1.5/100,000). In 1983 there was only one operation and since 1984 none.

Figure IV shows human operations for hydatids from 1971-1985.

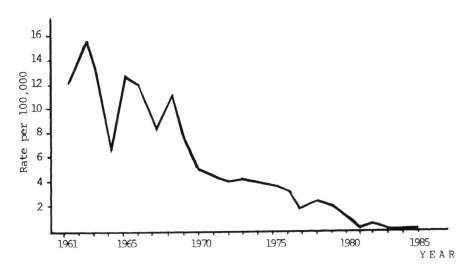

DISCUSSION

Apart from the great benefit accrued to the animal industry from the minimization of infection among food animals, as described above, the public health impact of the Campaign has been tremendous. It is estimated that from 1970 when the campaign was started up to 1985 about 750 operations were avoided. This can be considered as a real contribution of the Veterinary Department to public health protection in Cyprus.

The Anti-echinococcosis Campaign as originally planned, has within 15 years achieved its objectives. It has been officially terminated in 1985 and the situation will be monitored by carrying out selective dog testings, controlling stray dogs, keeping up education and maintaining abattoir control. These measures will ensure that echinococcosis will not become a problem again in the future.

Other island presently carrying out Echinococcosis control programmes are New Zealand, Tasmania (Australia) and the Falkland Islands. Iceland, has eradicated echinococcosis since the 1950s. New Zealand and Tasmania were first to use arecoline hydrobromide for dog-testing i.e. for the detection of E.granulosus taeniae in the faeces of dogs. The infected dogs in these two countries were not euthanised, as was the case in Cyprus, but were placed in quarantine and treated for deworming using arecoline hydrobromide or bunamidine hydrochloride. This slowed down progress and was a serious hazard to the dog owners. In the mid-1970s, with the discovery of praziquantel, a 100% taeniacide, the policy in these countries was revised and this drug was used combined with arecoline surveillance.

The policy of Cyprus for the deliberate destruction of stray,unwanted and infected dogs is correct as a control measure, but for various reasons this policy could not be applied in New Zealand and Tasmania. As a result, Cyprus was quicker in eliminating echinococcosis compared to those two countries, where Echinococcus control has been going on, on a national scale, since the early 1960s. The Falkland Islands are applying a highly-organized dog-dosing programme with praziquantel, and they are also achieving quick results, as evidenced from abattoir surveillance.

With the discovery of praziquantel, it is doubtful whether arecoline testing as used in New Zealand, Tasmania and Cyprus will ever be used again in new programmes. However, in any programme, effective dog control is essential so that new infections are avoided.

The factors of real importance in ensuring success of the Campaign in Cyprus were (i) effective control structure, (ii) strong educational component, (iii) efficient surveillance, (iv) effective slaughter control and (v) effective dog control.

The value of education and the role of man is to be pointed out in controlling this disease. This is increasingly recognized. Iceland,which was the first country to eradicate Echinococcosis places the main credit for this achievement upon education and enlightenment of its people about this disease.

REFERENCES

1. Polydorou, K. (1977): The anti-echinococcosis campaign in Cyprus.
    Tropical Animal Health Production, 141-146.
2. Polydorou, K. (1979): The problem of pets (dogs and cats) with special
    reference to the stray dog problem. 47th General Session of O.I.E.
    Committee, Rep. No. 500.
3. Polydorou, K. (1980): The control of echinococcosis in Cyprus. F.A.O.
    World Animal Review 33, 19-25.

68

TAENIASIS – THE TANTALISING TARGET

M.M.H. SEWELL

THE HOPES OF THE RECENT PAST

A few years ago it seemed possible, even probable, that recent developments in several fields would mean that effective, logistically feasible control of taeniasis-cysticercosis would be achieved within a relatively short time span. This expectation was largely based on work on *Echinococcus* or *Taenia saginata* but there seemed little reason to doubt that much could be applied to other taeniid cestodes, including *Taenia solium*.

These developments included: –

1. The degree of control that had been achieved over echinococcosis in Iceland and New Zealand using traditional methods and relatively inefficient taenicidal drugs. The methods which had been worked out in these countries were being or were about to be confidentally applied elsewhere.
2. The introduction of effective biochemical methods for distinguishing between the different species of the taeniid cestodes (28) that offered potentially more consistent differentiation than traditional morphological procedures, including those based on hard tissues (6).
3. The advent of anthelmintics, including the already somewhat dated niclosamide but also some of the benzimidazoles and praziquantel, which are highly effective at relatively small dose rates against the adult large taeniid tapeworms, and also of low toxicity with few or no side effects. It had even become possible to think in terms of mass treatment campaigns in man against *T. saginata*. Furthermore, some of these anthelmintics, notably praziquantel (48, 12) but also albendazol (5, 31), were known to be active against the metacestodes of *T. saginata*, while the former was soon to be shown to be active against those of *Taenia solium* in pigs (3) and man (41).
4. The demonstration by several groups of workers, well reviewed by Lloyd (36), that cattle can become resistant to challenge infections with oncospheres of *T. saginata* following natural infection or vaccination with living or crude preparations of oncospheres or even of the adult tapeworm and that the same applies with other taeniids used as laboratory models.
5. The adaptation to use with cysticercosis of relatively rapid and sensitive serological techniques, notably the enzyme-linked immunosorbent assay (ELISA) (22) and the rapid but much less sensitive counter-current immunoelectrophoresis (14, 39), which appeared to offer the possibility of effective *ante-mortem* diagnosis on-farm or in the abattoir.

PRESENT REALITIES

In the event, despite the clear guidelines laid down by FAO/UNDP/WHO (8), progress in controlling these parasites has been disappointingly slow. There is little evidence of major reductions in the prevalence of either *T. saginata* or *T. solium*. On the contrary it has become clear from recent work by Flisser (10) that the prevalence of c. 2% cysticercosis in human necropsies in Mexican hospitals reported at a conference in Mexico City in 1971 – and regarded at the time as somewhat alarmist — was accurate and has not significantly changed. Indeed this infection is probably of greater significance than was then realised, recent reports from China speaking of millions of infected pigs, with 12.5% of pigs in Jilin Province being infected in 1980, when there were also 150,000 cases of taeniasis in man and 10,000 cases of human cysticercosis in the same province (24). New foci of this infection are also still being reported in Africa (35).

At the same time there is evidence accruing that these parasites may not be as innocuous as was believed, at least in their intermediate hosts. This is because, in common with many other parasites, they appear able to suppress their host's immune response (38, 27, 11, 51). If this phenomenon results in their hosts being less able to control intercurrent disease or to respond adequately to vaccines, even non-zoonotic species such as *Taenia hydatigena* may be of greater general importance than was previously recognised.

Perhaps the most important of the difficulties in controlling taeniasis/cysticercosis is that defined by Gemmell (16) following a series of epidemiological studies in New Zealand (17). In that country echinococcosis had been very successfully controlled, yet the same methods failed to give a similar measure of control over the larger canine taeniids. This was despite the fact that most cestocidal drugs are more efficient against the large tapeworms than they are against *Echinococcus granulosus* and that modern drugs are more effective than those used during the greater part of the anti-hydatids campaign.

It is now realised that the large taeniids are so fecund that they tend to contaminate the environment much more heavily than *E. granulosus*. This occurs despite the fact that there are usually far fewer of the former than of *E. granulosus* in the typical final host. As a result the intermediate hosts experience a much greater and more constant infection by the large taeniids so that most, or even all, of them rapidly attain a very strong resistance to challenge by these parasites.

The introduction of control measures in such a hyperendemic situation may reduce the rate of contamination but this will merely result in the advent of a population of intermediate hosts in which many or all individuals do not have the previously invariable high resistance. Contamination of the environment continues because such control rarely, if ever, succeeds in eliminating all infection from the final hosts, whether man or dog, and large numbers of eggs will come from even a single tapeworm. Hence the now-susceptible, intermediate hosts are still at risk of infection, albeit in a relatively sporadic and geographically scattered manner. Not only do they then become infected but their susceptibility means that they are more liable to acquire heavy, readily detectable burdens — the "cysticercosis storm".

A good and long standing reflection of this can be seen in the difficulty which has been encountered in eliminating *T. saginata* from countries with highly developed public health facilities and the tendency to encounter cysticercosis storms in such areas if new developments in the disposal of human waste or in methods of animal management lead to an increased probability of infection (37).

The major practical problem associated with the identification of the large taeniids is the difficulty of telling their eggs apart or of distinguishing these from *Echinococcus* eggs, especially in epidemiological studies or during the investigation of the aetiology of cysticercosis storms. Serodiagnostic methods of identifying such eggs are being developed (4) but unfortunately there is rarely sufficient material available for use in biochemical methods of identification. This does not apply to the differentiation of the species and strains of *Echinococcus*, which is of great importance because of its relevance to the epidemiology of hydatid diseases in man. Accordingly most of recent studies on these methods of identifying taeniid tapeworms have concentrated on this (32, 33).

Despite encouraging reports from Russia, mass chemotherapy in man against such a mild a pathogen as *T. saginata* has not generally been found to be ethically acceptable. This is probably mainly because of the degree of official persuasion or even coercion that is necessary to ensure maximum coverage but also because of the risk of assumed, or even genuine, idiosynchratic reactions that may occur with any drug. Indeed, the very advent of effective and acceptable drugs against adult tapeworms, which can be safely dispensed on demand to individuals who complain of infection, may render the medical profession even less willing to support such an approach. It might still be possible to justify mass therapy against the more pathogenic *T. solium* if ultimate success in controlling the parasite in this way could be assured. However this would necessitate the effective prevention of re-infection, which would require the very dramatic improvement in food-hygiene which has already proved so elusive in areas where this infection is still prevalent. In the absence of such a comprehensive control scheme, similar factors to those which caused the difficulties in controlling the large tapeworms in dogs in New Zealand (16) would apply to *T. solium*. As regards the non-zoonotic tapeworms, the use of mass chemotherapy in dogs would probably be impossible on economic grounds, except — as in New Zealand — where it forms part of a campaign aimed primarily at hydatid disease in man.

The use of chemotherapy against the metacestode stage has continued to develop slowly. Praziquantel, in particular, is proving to be effective against cysticercosis, including neuro-cysticercosis, in man (44). However in many countries it remains too expensive to be used against cysticerci in human food animals, although this may not continue to be the case as patents run out and/or alternative sources or analogues become available, such as the Chinese pyquiton (52). Albendazole, which appeared to offer promise as a dual purpose drug already licenced for use in many countries as a nematocide in cattle and pigs, has not proved to be consistently effective against cysticerci (15, 45, 25). Finally, even when all the cysts have been killed by a suitable drug, the calcified residues, probably derived from those that were already dead or dying at the time of treatment, remain detectable for months (18) and might well bring this approach into disrepute.

For immunoprophylaxis the problem has not lain so much in the difficulty of developing an effective vaccine as in devising means of producing such a vaccine in adequate quantities and at an economic cost. An inexpensive vaccine would be essential for use against such a mild zoonosis as *T. saginata* that only occurs sporadically in countries with highly developed public hygiene systems and is hyperendemic in other areas. Furthermore, the commercial production of a vaccine which had to be made from parasitic material derived from mammals — especially man — would surely present insuperable logistic problems.

Serodiagnosis of cysticercosis caused by *T. solium* in man may be of epidemiological value (26), although even here the relatively frequent occurrence of false positive and false negative reactions still causes considerable difficulties in interpretation. For the investigation of individual suspected cases of neuro-cysticercosis in man, the detection of specific antibodies in cerebro-spinal fluid is both reliable and effective (7, 36, 42) although serodiagnosis is probably best used alongside other proven diagnostic methods, such as computerised tomography (2).

In domestic animals the difficulties associated with sero-diagnosis are more severe (49, 53). The problem appears three-fold. Firstly the host reaction to the parasite is relatively weak, perhaps partly as a result of immunosuppression by the parasite. Secondly the humoral response is complex and some of the antibodies may cross-react with antigens derived from a variety of other sources. Thirdly the continued production of the antigens by the parasite leads to a situation in which the serum could contain at the same time a small excess of any particular antigen or antibody or antigen-antibody complexes. Hence a simple increase in the sensitivity of a test may result in more false positive reactions, while conversely an attempt to increase the specificity of the test tends to result in reduced sensitivity — and this in a complex system where the concentrations of several antigens and antibodies may vary independently.

A further difficulty, which even accurate sero-diagnosis will encounter if it is to be used in the control of bovine cysticercosis, arises from the fact that some of the apparent false-negatives are probably more a reflection of the inaccuracies long known to be inherent in the direct knife-and-eye methods for detecting cysticerci (34, 23) than of any defect in the serological procedure. It is almost certain that an accurate diagnostic procedure would greatly increase the number of infected carcasses that are detected (50) but the cyst(s) involved would often prove difficult to find without unacceptably expensive dissection of the carcass. Such an increase in the detection rate of bovine cysticercosis is at least as likely to lead to the butchers, owners and politicians insisting on a return to less effective detection methods as it is to an attempt to reduce the prevalence of infection! The probable response would be an attempt to discredit even a truly accurate sero-diagnostic procedure.

FUTURE PROSPECTS

Our greater understanding of the epidemiology of these infections and of the problems associated with their control offers opportunities for using modern technology in attempting to develop a useable new approach to their control.

Firstly, the proposition can now be accepted and defended on scientific, epidemiological grounds that traditional methods for controlling these infections are not likely to eliminate them even in areas where both the official will and funds are available to provide the necessary general motivation and facilities. This is especially true for *T. saginata* but may also be true for *T. solium* in those countries where recent changes in the terms of international trade, such as the price of oil, have rendered the necessary improvements in educational and hygienic facilities improbable, at least in the short term.

This should facilitate the task of seeking funding for research into alternative approaches to control and their application, since it is now easier to justify the need for such alternatives on both economic and public-health grounds.

The requirements for an effective alternative approach to the control of human tapeworms are clear. It must be based on controlling the infection in the intermediate host by relatively inexpensive means, which are applicable to all the animals at risk and are potentially able to virtually eliminate the parasites without lessening the hosts' resistance to any residual natural challenge.

It is now possible to envisage how such a solution could be attained and to define the research and administrative targets that are necessary if it is to be achieved. However an essential prerequisite will be to persuade funding bodies that these targets are attainable and that the end is worth the application of some of their invariably scarce resources.

In the first place there is a continuing need for a larvicidal cestocide which can be produced at a price at which it could be widely used, at least in young stock. For preference, total reliance should not be placed in any one drug, such as praziquantel, even when legal and technological developments result in this becoming less costly. The recent observations that flubendazole is effective against the metacestodes of *T. solium* in pigs (46) and man (47) and that high doses of fenbendazole are effective against immature and mature cysticerci of *T. saginata* (1) and the new Chinese metacestocide "Mienang No. 5" (24) merit further investigation.

72

However, if the drug firms are to continue with this research they need to be confident that there will be a remunerative market for their products. It is unlikely that such a market will exist if it is based solely on individual action by animal owners in a situation where many infected animals escape detection and treated animals are liable to become reinfected. As the epidemiological status moves from hyperendemic to endemic such re-infection becomes more likely. Accordingly it will be essential for metacestocidal drugs to be used in conjunction with an effective and cost-effective vaccination procedure that will prevent re-infection. It is difficult to see how this could be logistically feasible unless the vaccine is non-living, so that it can be readily incorporated into another vaccine used in the same young stock.

It will probably also be necessary to greatly improve the accuracy of detection of infected animals or carcasses and it is here that the relevance of sero-diagnosis becomes apparent. In particular the development of highly specific and sensitive procedures for detecting the characteristic products of metabolism of viable cysts in the circulation would be very advantageous. This would allow the identification of the young animals or groups of animals which require treatment and also the detection of those slaughter-weight animals which present a genuine hazard to human health. The exclusion of animals which only contain dead cysts that no longer produce these metabolic products should be a major factor in rendering the test acceptable to owners, butchers and public health authorities. Diagnostic procedures based on the use of monoclonal antibodies and the antigens that can be isolated by using these as immuno-affinity reagents could play a vital role in allowing the development of tests which have very low backgrounds, so that their sensitivity can be increased without incurring an unacceptable loss of specificity, perhaps by the use of cocktails of monoclonal antibody derived reagents.

Two areas of research are vital precursors for both these applications. These are, firstly the elucidation of the immunochemistry of the antigens and immunogens derived from *T. solium, T. saginata* and the other taeniid cestodes (19) and secondly immunogenetic studies such as those recently described in a model system by Lightowlers *et al* (43) and now being extended to parasites of veterinary importance (9).

The approach adopted by Harrison and Parkhouse (20, 21), in which they seek to identify the targets of the protective immune response at the molecular level by the use of protective monoclonal antibodies against the larval stages of *T. saginata* provides a logical sequence designed to produce a protective immunogen by genetic engineering (13). However, Simpson *et al* (43) caution that this may yet present many practical difficulties.

Such an integrated alternative approach to the control of these parasites will only be developed and tested and, of equal importance, the results will only be applied if both the public health authorities and their political paymasters are prepared to support this by funding the research and by introducing measures which will both penalise the careless owner, whose unprotected animals are found to be infected, and at the same time indemnify those who take the prescribed control measures but still have animals which the *ante-mortem* tests show to be infected. Such an even-handed carrot-and-stick approach should be far more acceptable to owners than the present haphazard detection procedure which relies almost entirely on penalising the unfortunate. At the same time, provided really effective control measures and accurate detection procedures can be developed, the cost to government would be minimal.

1. Bessonov AS, Arkipova NS, Ubiraer SP, Uspenski AN and Shekhosvotvo NV: Trudy Vsesoyuznogo Instituta Gel'mintologii im KO Skrjabina. 27: 13–17, 1984
2. Byrd SE, Daryabagy J, Thompson R, Zant J, Locke GE and Biggers S: The computed tomographic spectrum of cerebral cysticercosis. Journal of the National Medical Association, USA. 77: 553–560, 1985.

3. Chavarria M and Gonzalez D: Droncit en el tratimento de la cisticercosis porcina. Esp. Vet. 1: 160–165, 1979.

4. Craig PS: Immunodifferentiation between eggs of *Taenia hydatigena* and *T. pisiformis*. Annals of Tropical Medicine and Parasitology. 77: 537–538, 1983.

5. Craig TM and Ronald NC: Preliminary studies on the effect of albendazole on the cysticerci of *Taenia saginata*. Southwestern Veterinarian. 31: 121–124, 1978.

6. Edwards GT and Herbert IV: Some quantitative characters used in the identification of *Taenia hydatigena, T ovis, T pisiformis,* and *T multiceps* metacestodes. Journal of Helminthology. 55: 1–7, 1981.

7. Estrada JJ and Kihn RE: Immunochemical detection of antigens of larval *Taenia solium* and anti-larval antibodies in the cerebrospinal fluid of patients with neurocysticercosis. Journal of the Neurological Sciences. 71: 39–48, 1985.

8. FAO/UNDP/WHO: Guidelines on surveillance, prevention and control of taeniasis/cysticercosis. Document VPH/84 49. World Health Organisation, Geneva, Switzerland. 1984.

9. Fischer M and Howell MJ: Isolation, cloning and characterization of different genes in *Taenia ovis*. Immunobiology and Molecular Biology of Cestode Infections. Coopers Animal Health Symposium, Melbourne. A7, 1986.

10. Flisser A: Cysticercosis: a major threat to human health and livestock production. Food Technology. 39: 61–64, 1985.

11. Flisser A, Aluja A, Rodriguez-Carbajal J, Gonzalez D, Correa D, Skurovich MA, Plancarte A, Tovar A, Garcia AM, Ostokovsky P, Montero R, Madrazo I, Stefano A, Rodriguez-del-Rosal E, Cohen S and Fernandez B: Porcine cysticercosis treated with praziquantel. A controlled multidisciplinary study. Immunobiology and Molecular Biology of Cestode Infections. Coopers Animal Health Symposium. Melbourne. D9, 1986.

12. Gallie GJ and Sewell MMH: The efficacy of praziquantel against the cysticerci of *Taenia saginata* in calves. Tropical Animal Health and Production. 10: 36–38, 1978.

13. Gamble HR and Zarlenga DS: Biotechnology in the development of vaccines for animal parasites. Veterinary Parasitology. 20: 237–250, 1986.

14. Geerts S, Kumar V, Aerts N and Ceulemans F: Comparative evaluation of immuno-electrophoresis, counterimmunoelectrophoresis and enzyme-linked immunosorbent assay for the diagnosis of *Taenia saginata* cysticercosis. Veterinary Parasitology. 8: 299–307, 1981.

15. Geerts S and Kumar V: The effect of albendazole on *Taenia saginata* cysticerci. Veterinary Record. 109: 217, 1981.

16. Gemmell MA: A critical approach to the concepts of control and eradication of echinococcosis/hydatidosis and taeniasis/cysticercosis. Parasitology. Quo Vadit? Proceedings of the Sixth International Congress of Parasitology. Canberra, Australia. 465–472, 1986.

17. Gemmell MA and Johnstone PD: Experimental epidemiology of hydatidosis and cysticercosis. Advances in Parasitology. 15: 311–369. 1977.

18. Harrison LJS, Gallie GJ and Sewell MMH: Absorbtion of cysticerci in cattle after treatment of *Taenia saginata* cysticercosis with praziquantel. Research in Veterinary Science. 37: 378–380, 1984.

19. Harrison LJS and Parkhouse M: Antigens of taeniid cestodes in protection, diagnosis and escape. Current Topics in Microbiology and Immunology. 120: 159–168, 1985.

20. Harrison LJS and Parkhouse M: Passive protection against *Taenia saginata* infection in cattle by a murine monoclonal antibody reactive with the surface of the invasive oncosphere. Parasite Immunology. 8: 319–332, 1986a.

74

21. Harrison LJS and Parkhouse M: Identification of protective antigens in *Taenia saginata* cysticercosis. Immunobiology and Molecular Biology of Cestode Infections. Coopers Animal Health Symposium. Melbourne, A2, 1986b.

22. Harrison LJS and Sewell MMH: A comparison of the enzyme-linked immunosorbent assay and the indirect haemagglutination technique applied to sera from cattle experimentally infected with *Taenia saginata* (Goeze, 1782). Veterinary Immunology and Immunopathology. 2: 67–73, 1981.

23. Heath DD, Lawrence SB and Twaalhoven H: *Taenia ovis* cysts in lamb meat: the relationship between the numbers of cysts observed at meat inspection and the numbers of cysts found by fine slicing of tissue. New Zealand Veterinary Journal. 33: 152–154, 1985.

24. Jing JS and Wang PY: Advance of studies on cysticercosis of man and pigs at home and abroad in recent years. Chinese Journal of Veterinary Science and Technology (Zhongyo Shouyi Keji). No. 6: 33–36, 1985.

25. Kassai T, Takats C, Rendl, P and Fok E: Effect of albendazole and praziquantel on *Taenia saginata* cysticerci. Helminthologia. 21: 295–302, 1984.

26. Larralde C, Laclette JP, Bojelli R, Diaz ML, Goverensky T, Montoya RM and Goodssaid F: Toward a reliable serology of *Taenia solium* cysticercosis with vesicular fluid antigens. Immunobiology and Molecular Biology of Cestode Infections. Coopers Animal Health Symposium. Melbourne, E4, 1986.

27. Leid RW and McConnell LA: PGE_2 generation and release by the larval stage of the cestode *Taenia taeniaeformis*. Prostaglandins, Leukotrienes and Medicine. 11: 317–323, 1983.

28. LeRiche PD and Sewell MMH: Differentiation of taeniid cestodes by enzyme electrophoresis. International Journal for Parasitology. 8: 479–483, 1978.

29. Lightowlers MW, Rickard MD and Mitchell GF: Development of a model vaccine against cysticercosis. Australian Veterinary Journal. 62 (Supplement): 5, 19

30. Lloyd S: Passive and active immunisation against cysticercosis. Some important parasitic infections in bovines considered from economic and social (zoonosis) points of view. Parasitological Symposium. Lyons, 26–26 October 1983. Eds J Euzeby and J Gevrey. 187–197.

31. Lloyd S, Soulsby EJL and Theodorides VJ: Effect of albdendazole on the metacestodes of *Taenia saginata* in calves. Experimentia. 34: 723–724, 1978.

32. McManus DP and McPherson CNL: Strain characterisation in the hydatid organism, *Echinococcus granulosus*: current status and new perspectives. Annals of Tropical Medicine and Parasitology. 78: 193–198, 1984.

33. McManus DP and Simpson JG: Identification of the *Echinococcus* (Hydatid disease) organisms using cloned DNA markers. Molecular and Biochemical Parasitology. 17: 171–178, 1985.

34. Mann I and Mann E: The distribution of measles (*Cysticercus bovis*) in African bovine carcasses. The Veterinary Journal. 103: 239–251, 1947.

35. Marty P, Hertzog U, Marty-Janssan L, LeFichoux Y and Douchet J: Deux cas de cysticercose observes au Cameroun. Medecine Tropicale. 45: 83–86, 1985.

36. Mohammad IN, Heiner DC, Miller BL, Goldberg MA and Kaga IG: Enzyme-linked immunosorbent assay for the diagnosis of cerebral cysticercosis. Journal of Clinical Microbiology. 20: 775–779, 1985.

37. Nansen P: The epidemiology of bovine cysticercosis (*C. bovis*) in relation to sewage and sludge application on farmland. Epidemiological Studies of Risks Associated with the Agricultural Use of Sewage Sludge: Knowledge and Needs. Eds JC Block, AH Havelaar and PL Hermite. London and New York: Elsevier Applied Science Publishers, 76–82, 1986.

38. Nichol CP and Sewell MMH: Immunosuppression by larval cestodes of *Babesia microti* infections. Annals of Tropical Medicine and Parasitology. 78: 228–233. 1984.

39. Pathak KML, Gaur SNS and Garg SK: Counter current electrophoresis, a new technique for the rapid diagnosis of porcine cysticercosis. Journal of Helminthology. 58: 321–324, 1984.

40. Rhoads ML, Murrel KD, Dilling GW, Wong MM and Baker NF: A potential diagnostic agent for bovine cysticercosis. Journal of Parasitology. 71: 779–787, 1985.

41. Robles C and Chavarria MC: Presentation de un caso clinico de cisticercosis cerebral tratado medicamente con un nuevo farmaco: praziquantel. Salud Publica Mexico. 21: 603–618, 1979.

42. Rosas N, Sotelo J and Nieto D: ELISA in the diagnosis of neurocysticercosis. Archives of Neurology. 43: 353–356, 1986.

43. Simpson AJG, Walker T and Terry R: An introduction to recombinant DNA technology. Parasitology. 91: S7–S14, 1986.

44. Sotelo J, Torres B, Rubio-Donnadieu F, Escobedo F and Rodriguez-Carabajal J: Praziquantel in the treatment of neurocysticercosis: long term follow-up. Neurology. 35: 752–755, 1985.

45. Stevenson P, Holmes PW and Muturi JM: Effect of albendazole on *Taenia saginata* cysticerci in naturally infected cattle. Veterinary Record. 109: 82, 1981.

46. Tellez-Giron E, Ramos MC, Dufour L, Montante M, Tellez E Jr, Rodriguez J, Mendez FG and Mireles E: Treatment of neurocysticercosis with flubendazole. American Journal of Tropical Medicine and Hygiene. 33: 627–631, 1984.

47. Tellez-Giron E, Dufour L and Montante M: Effect of flubendazole on *Cysticercus cellulosae* in pigs. American Journal of Tropical Medicine and Hygiene. 30: 135–138, 1981.

48. Thomas H and Gonnert R: The efficacy of praziquantel against experimental cysticercosis and hydatidosis. Zeitschrift fur Parasitenkunde. 55: 165–179, 1978.

49. Urquhart GM: Immunodiagnosis of bovine cysticercosis. Some important parasitic infections in bovines considered from economic and social (zoonosis) points of view. Parasitological Symposium. Lyons, 24–26 October 1983. Eds J Euzeby and J Gevrey. 183–186, 1984.

50. VanKnapen F, VanDerLugt G and Franchimont JH: Immunodiagnosis of *Taenia saginata* cysticercosis. Some important parasitic infections in bovines considered from economic and social (zoonosis) points of view. Parasitological Symposium. Lyons, 24–26 October 1983. Eds J Euzeby and J Gebrey, 1984.

51. WesLeid R, Suquet CM, Grant RF, Tanigoshi L, Blanchard DB and Vilma T: Taeanistatin, a cestode proteinase inhibitor with broad host regulatory acitivty. Immunobiology and Molecular Biology of Cestode Infections. Coopers Animal Health Symposium. Melbourne, C9, 1986.

52. Xu ZB, Chan MP, Hung ML and Chung WN: Efficacy of pyquiton (praziquantel) in treatment of human cysticercosis. Beijing Medical Journal. 4: 326–328, 1982.

53. Yong WK, Heath DD and VanKnapen F: Comparison of cestode antigens in an enzyme-linked immunosorbent assay for the diagnosis of *Echinococcus granulosa, Taenia hydatigena,* and *T. ovis* infections in sheep. Research in Veterinary Science. 36: 24–31, 1984.

OBSERVATIONS ON POSSIBLE STRAIN DIFFERENCES IN TAENIA SAGINATA

G.WOUTERS, J.BRANDT and S.GEERTS

ABSTRACT
   Strain differences are a well known phenomenon in several taeniid
species. Up to now, however, almost no mention has been made on the existence of strains of Taenia saginata.
   Results obtained by enzyme electrophoresis and experimental infections
in Meriones unguiculatus as an alternative intermediate host indicate the
possibility of strain variation in T.saginata. An enzyme electrophoretic
study of 30 adult T.saginata worms for hexokinase showed that there were
3 distinct zymogram patterns. After subcutaneous inoculation of M.unguicu-
latus with 24 batches of T.saginata oncospheres, it was observed that some
batches were significantly more infective for these jirds than others.
Data from literature on the other hand prove that there might exist
T.saginata strains in Africa (Nigeria and Kenya) where the metacestodes
seem to occur preferably in the liver rather than in the muscles. In Taiwan
there is some evidence that a T.saginata strain exists where intermediate
hosts other than cattle are involved. A similar situation might exist in
the Phillipines. All these data strongly suggest that - similarly to other
Taenia spp - strain differences do exist in T.saginata.

1. INTRODUCTION
   Strain differences in several taeniid species are a well known phenomenon
in several taeniid species e.g. Echinococcus granulosus, Taenia taeniaeformis, Taenia crassiceps etc.With the exception of Abuladze et al.(1),
however, up to now no mention has been made in the literature on the existence of different strains within the Taenia saginata species. In the
present study data from experiments and indications in literature are analysed in view of T.saginata strain variations.

2. MATERIAL AND METHODS
2.1. Enzyme electrophoresis:
   T.saginata proglottids from Belgium, Burundi and Zaïre were used. Electrophoresis was carried out using the mini-slab electrophoresis system 1001
(Idea Scientific Company).
- Preparation of the samples: proglottids were washed in physiological
saline and homoginised in an iso-osmotic enzymestabilisator (50% v/v) and
after ultracentrifugation (80.000g; 30 min) the supernatant was stored
at -20°C in small aliquots. After storage 20 ul was added to 40 ul of a
sucrose solution (2g/l).
- Electrophoresis: preparation of acrylamide gels: to 10.4 ml of distilled
water 8 mg ammonium persulphate, 0.55 ml of carrier ampholytes (40%, LKB,
pH range 3-10), 2.5 ml of an acrylamide stock solution (400 g acrylamide and

10 g bisacrylamide per liter) and 10 µl N,N,N',N'-tetramethylethylenedia-
mine (TEMED) were added. This gel solution was poured in a frame and poly-
merised in 30 minutes. Gels were 0.5 mm tick. A vertical system is used
with two buffercompartiments: 0.02 mol/l H_3PO_4 at the anode and 0.02 mol/l
NaOH at the cathode. 10 µl of the sample is applied in each slot of the
cathodal side.
- A constant voltage for 2.5h with an initial current of 12 mA.
- Staining: After electrophoresis the gel was removed from one glassplate
and left in a developing overlay.
Phosphoglucose-isomerase (PGI: EC.5.3.1.9): 4.5 ml aqua dest., 1.2ml 0.4 M
Tris pH 8, 5 mg NADP, 10 µl glucose-6-phosphate dehydrogenase (G6PDH), 5mg
Fructose-6-phosphate and 1ml 0.1 M $MgCl_2$.
Hexokinase (HK: EC. 2.7.1.1.): 5.5ml 0.1 M Tris pH 7.4, 3mg NADP, 10 ul
G6PDH, 250 mg glucose, 50mg ATP, 1ml 0.1 M $MgCl_2$, 2.5mg KCN.
Each solution contains 1.5ml 3-(4,5-dimethyl-2-thiazolyl) 2,5-diphenyl-
2N-tetrazoliumbromide)(MTT) and 30 µl Meldola Blue (6g/1). This solution was
added to 5ml agarose (200g/1, 60°C). Gels were then kept at 37°C (10-20
minutes).

2.2. Inoculation of T.saginata oncospheres
In total 387 jirds (Meriones unguiculatus) were inoculated subcutaneous-
ly with 24 different batches of hatched and activated T.saginata oncosphe-
res (Tables 1, 2 and 3).
The animals were given immunosuppressive treatments as described by
Wouters et al.(11).

3. RESULTS
3.1. Enzyme electrophoresis:
Fig.1 shows the zymogram of hexokinase for 30 individual T.saginata.
There is only one PGI pattern observed. The first HK-pattern was most common, whereas the second HK-pattern was only seen three times. There was no
relation with geographical origin; the three HK-patterns occured in Belgian
and Burundese samples.

Fig.1: Zymogram for HK of T.saginata (adult worms).

3.2. Inoculation of T.saginata oncospheres in M.unguiculatus.
The results of the experimental infections are summarised in Tables 1, 2
and 3. As far as possible experimental conditions were kept the same
although the dose, activation percentage and age of the eggs were not always identical. None of these factors, however, had a significant effect on
the percentage of the infected animals (11). Four batches (3, 11, 14, 19)

gave a higher infection rate than the others. The numbers 3, 14 and 19 differ significantly from all the others as calculated by variance analysis (Duncan's multiple range test).

Table 1: Subcutaneous inoculation of dexamethasone treated M.unguiculatus with T.saginata oncospheres of Belgian origin.

T.saginata batch number	2	3	4	5	6	7		
No. of inoculated animals	7	6	4	32	40	16	5	5
Animals with developing metacestodes								
Number	1	3	1	1	2	0	0	0
%	14.3	50	25	3.1	5	-	-	-

Table 2: Subcutaneous inoculation of splenectomised M.unguiculatus with T.saginata oncospheres of Belgian origin.

T.saginata batch number	8	9	10	11	12	13	14		15	16	17
No. of inoculated animals	9	10	10	10	4	6	5	3	8	5	5
Animals with developing metacestodes											
Number	0	0	0	3	0	0	1	1	0	0	0
%	-	-	-	30	-	-	20	33	-	-	-

Table 3: Subcutaneous inoculation of M.unguiculatus with T.saginata oncospheres of African origin.

T.saginata batch number	18			19			20	21	22	23	24	
Origin	Bu			Bu			Bu	Bu	Bu	Bu	Sn	
Immunosuppressive treatm.	Dx	Sp	N	Dx	Sp	N	Dx	Dx	Dx	Dx	Dx	N
No. of inoculated animals	10	4	6	10	10	5	5	5	5	5	43	20
Animals with developing metacestodes												
Number	0	0	1	1	5	0	0	0	0	0	0;	0
%	-	-	17	10	50	-	-	-	-	-	-	-

(Dx: dexamethasone, Sp: splenectomy, N: no immunosuppressive treatment, Bu: Burundi, Sn: Senegal).

4. DISCUSSION

The results obtained by enzyme-electrophoresis and experimental infections of M.unguiculatus indicate the possibility of strain variation in T.saginata. Although for T.taeniaeformis there is a correlation between the intermediate host (rat, mouse) and the HK-pattern, there is no clear evidence in the relation between strains and HK-patterns for T.saginata. Strains of T.taeniaeformis were also identified with HK isoenzymes (Wouters, unpublished data). The results of the infectivity tests of M.unguiculatus have also to be interpreted with caution since the animals were not inbred, so that there could be a substantial variation in the susceptibility of the individual animals. Since a high number of animals were used, however, and since all the animals came from the same breeding unit, the significant differences in infectivity from batch to batch are strongly suggestive for strain differences. Up to now no reports have been published on strain variation in T.saginata, except by Abuladze et al. (.1) . . After experimental infections by calves these authors observed a longer life span and a higher survival rate of cysticerci derived from tapeworms from southern areas of the USSR than those from northern and central areas. In the literature however, several authors working in Africa mention the striking presence of Cysticercus bovis infections in the liver of cattle. Buck et al.(2) already described this in 1935. Reports from Nigeria (Belino (3) and Schillhorn Van Veen (9)), Kenya (Ginsberg and Grieve (7))and Sudan (El Sadik Abdullah (5))demonstrated that living cysticerci in these countries are very common in the liver. Several cases are described where the cysticerci could be found in the liver only, even after meticulous search for cysts in the other organs and the muscles. Since the cysts had no hooks they were identified as Cysticercus bovis. Only Schillhorn Van Veen (9) suggested T. saginata var.giraffae as a possible explanation for the cysticerci in the liver, since it was observed by Price (8) that liver infections occured in calves which were challenged with embryos originating from unarmed liver cysticerci of a giraffe (after human passage). These findings have to be reexamined in the light of recent reports of Wong et al. (10) and Fan et al.(6) on a Taiwanese Taenia, with the adult morphology of T.saginata but the metacestodes of which were found only in the liver and possessed rudimentary rostellar hooklets. Goats, pigs and calves could be infected experimentally with this species. It was not yet possible to obtain material from liver cysticerci from Africa or Taiwan to compare this with the 'normal'T.saginata . Equally the presence of rudimentary hooks in the African liver cysticerci might have been overlooked and certainly their potential development in alternative hosts remains to be studied. The Philippine Taenia, as described by Arambulo et al. (2), might also belong to the same strain or (sub)species as the Taiwanese one.

From all these data, as well from literature as from experimental work, it can be   concluded that the existence of different strains of T.saginata is very likely.

REFERENCES

1. Abuladze KI, Gil'denblat AA and Buslaeva TP: The contemporary opinion on 'strains' and new data on ultrastructure of C.bovis.Zakavkazskoi Konferentsii po obshcei Parasitologii Maya: 50-58, 1977.

2. Arambulo PV, Cabrera BD and Tongson MS: Studies on the zoonotic cycle of Taenia saginata taeniasis and cysticercosis in the Philippines. International Journal of Zoonoses, 3: 77-104, 1976.

3. Belino ED: Some observations on Taenia saginata cysticercosis in slaughter cattle in Nigeria. International Journal of Zoonoses, 2: 92-99, 1975.

4. Buck, Lambertin and Belona R: Localisation hépatique de Cysticercus bovis et Cysticercus cellulosae. Receuil de Médecine Vétérinaire, 8: 74, 1935.

5. El Sadik A: Distribution of bovine cysticercosis in Sudan from an abattoir survey in Khartoum.Sbornik Nauchn. Trudov Medical Veterinary Akademy, 108: 120-122, 1979.

6. Fan PC, Chan CH, Chung WC, Wong MM, Chen CC and Wu CC: Taiwan Taenia may be considered as a new species of Taenia. Abstracts of the Sixth International Congress of Parasitology 1986. Brisbane, Australia. (Handbook supplement, p.10).

7. Ginsberg A and Grieve IM:Two unusual cases of liver cysticercosis. The Veterinary Record, 71: 618, 1959.

8. Price EV: A note on hepatic cysticercosis, with the proposal of a new variety of Taenia saginata. Journal of the Alabama Academy of Science, 32: 257, 1961.

9. Schillhorn Van Veen TW: The occurrence of Cysticercosis bovis in cattle livers. The Veterinary Record, 104: 370

10. Wong MM, Chen CC, Chung WC, Chao D and Fan PC: Taiwan aboriginal taeniasis-an intermediate form? Abstracts of the Sixth International Congress of Parasitology 1986, Brisbane, Australia (p259).

11. Wouters G, Geerts S, Brandt J, Kumar V and Calus A: Meriones unguiculatus as an experimental host for Taenia saginata metacestodes. International Journal of Parasitology (submitted).

IMMUNO-PROPHYLAXIS OF *TAENIA SAGINATA* CYSTICERCOSIS

L.J.S. HARRISON and G.W.P. JOSHUA

ABSTRACT

There are several conventional approaches which can be taken in the control of *Taenia saginata* cysticercosis in cattle and the adult tapeworm infection in man. However, where such measures have been applied, they have tended to reduce the incidence but not eradicate the parasite. Additional control methods such as immuno-prophylaxis and chemotherapy are needed. Production of a vaccine may be feasible since the acquired immunity displayed by cattle in the field can be stimulated artificially by vaccination with extracts of the parasite. Recent studies have been directed at the identification of the molecular 'targets' of this protective immune response in cattle. The identification of such 'targets' should facilitate the production of a vaccine by modern molecular biological techniques.

INTRODUCTION

Taenia saginata is a man/cattle taeniid of cosmopolitan distribution and of economic, veterinary and public health importance. The World Health Organisation have published guidelines for its control (8). In the developed countries, conventional methods, such as improved sanitation, drug treatment of infected humans and meat inspection, have reduced the incidence, but not eradicated the parasite. In less developed countries, with poor sanitation, the incidence remains high despite efforts of control, for instance, at meat inspection. The resulting condemnation of heavily infected carcasses or downgrading the value of lightly infected carcasses can cause unacceptable economic losses in countries with a heavy investment in the beef industry.

Gemmell, in a recent review of Taeniasis/cysticercosis control (9), concluded that the difficulty in eradicating species such as *T. saginata*, may lie in their high biotic potential. In addition, in high incidence areas cattle rapidly develop immunity to reinfection due to ingestion of eggs from pasture (24). Consequently the numbers of cysticerci in the cattle may be lower than might be expected from the high infection pressure. If the numbers of eggs on the pasture drops, cattle may lose their natural immunity and become susceptible to sporadic challenge. This is the situation which occurs in low incidence areas such as Britain where cattle in localised outbreaks can become heavily infected. Similarly, in New Zealand, application of conventional control measures designed to eliminate hydatid disease, resulted in an increase in the incidence of *Taenia ovis* infection in sheep. *T. ovis* is considered to have a biotic potential similar to that of *T. saginata*. The concern is that the appplication of conventional control measures to high incidence areas of *T. saginata* may result in an overall increase in the numbers of cysticerci in cattle. A vaccine is, therefore, needed to replace natural immunity until the parasite is eradicated. This would apply even if chemotherapy was introduced as an additional control measure in cattle (14).

The purpose of this paper is to briefly review progress towards the production of such a vaccine against *T. saginata* infection in cattle.

IMMUNITY TO *T. SAGINATA* CYSTICERCOSIS

Development of a vaccine may be feasible as there is evidence that *T. saginata* infection in cattle can be eliminated by immune mechanisms. Resistance to secondary infection can be produced either by primary infection with the parasite (4, 5, 7) or by immunisation using parasite extracts (6, 19, 22). The oncosphere, or material secreted or shed by the oncosphere, is thought to be one potent source of immunogens. However, until recently, little work has been done on the characterisation at the molecular level of the antigens involved (11).

A considerable amount of evidence points to the importance of humoral immunity in resistance to the early stages of taeniid parasites. Protection by passive transfer of immune sera in systems such as *Taenia taeniaeformis* indicated that antibodies play a definitive role in immunity and that it is the invasive larvae or oncosphere which are vulnerable to such attack (23). Protection from *T. saginata* infection can similarly be transferred to neonatal cattle (20).

More recently a Western blotting study on *T. saginata* oncospheres in conjunction with sera from resistant and susceptible cattle identified certain oncospheral components which were recognised by antibodies from resistant animals. Balb/C mice responded to the same antigens so the production of monoclonal antibodies directed against these determinants is feasible (15).

An analysis of the surface, excretory/secretory and somatic compartments of *T. saginata* oncospheres by a combination of direct and biosynthetic radio-labelling and SDS-PAGE revealed a restricted number of components in all three compartments. Mouse derived monoclonal antibodies directed against the oncospheral stage demonstrated both stage specific and common determinants on the surface of oncospheres and metacestodes. One of these monoclonal antibodies, when injected into cattle conferred protection against oral challenge with *T. saginata* eggs (12). Recent studies indicate that the antigenic component identified by this monoclonal, once purified from an extract of whole oncospheres, may be involved in stimulating immunity in cattle to oral challenge with the eggs (13).

Although the invasive oncosphere is considered the prime 'target' of the immune response, components of the metacestode stage may also constitute possible 'targets'. There would be clear advantages in having a vaccine which was directed against both the early invasive and later developmental stages of the parasite. If any invasive parasites which initially survived they could then be killed at a later stage in development.

As shown by the monoclonal antibodies raised against *T. saginata* oncospheres (12), some components may be shared between oncospheres and metacestodes. This also appears to be the case with *T. taeniaeformis*. However, other metacestode components may be unique (10). Recent studies into the surface, excretory/secretory, somatic and cyst fluid compartments of *T. saginata* so far indicate that an antibody response to certain metacestode components may indeed be associated with immunity (17, 21). It is worth noting that most work up till now has concentrated either on invasive oncospheres or on the fully developed metacestodes. Intermediate developmental forms now require study and this opinion has been expressed by other authors (1, 16).

As yet there are no reports in the literature of the application of molecular biology in the search for a vaccine against *T. saginata*. However, progress in this field is being made with other taeniid species such as *T. taeniaeformis* (2, 18) and *T. ovis* (3). Clearly extending the application of these techniques to parasites such as *T. saginata* could do much to overcome the problems encountered when trying to produce vaccines against a parasite, where the main source of parasitic material is man or cattle.

CONCLUSIONS

The last few years have seen the start of the application of modern analytical/molecular biological approaches to taeniids, including *T. saginata*. Hopefully, these approaches should open up new possibilities for control by immuno-prophylaxis. Parasites such as *T. saginata*, which display strong evidence that they can stimulate an effective immunity in their bovine hosts, must surely be suitable candidates for further study.

ACKNOWLEDGEMENTS
Work in this laboratory is presently supported by grants from the Overseas Development Administration of the Foreign and Commonwealth Office and the Wellcome Trust.

REFERENCES

1. Bogh HO, Rickard MD and Lightowlers MW: Stage specific immunity following vaccination against *Taenia taeniaeformis* infection in mice. Immunobiology and Molecular Biology of Cestode Infections. Coopers Animal Health Symposium. Melbourne. A5, 1986.
2. Bowtell DDL, Saint RB, Rickard MD and Mitchell FG: Expression of *Taenia taeniaeformis* antigens in *Escherichia coli*. Molecular and Biochemical Parasitology. 13: 173−185, 1984.
3. Fisher M and Howell, MJ: Isolation, cloning and characterisation of different genes in *Taenia ovis*. Immunobiology and Molecular Biology of Cestode Infections. Melbourne. A7, 1986.
4. Gallie GJ and Sewell MMH: The survival of *Cysticercus bovis* in resistant calves. Veterinary Record. 91: 481−482, 1972.
5. Gallie GJ and Sewell MMH: The serological response of three month old calves to infection with *Taenia saginata (Cysticercus bovis)* and their resistance to reinfection. Tropical Animal Health and Production. 6: 163−171, 1974.
6. Gallie, GJ and Sewell MMH: Experimental immunisation of six month old calves against infection with the cysticercus stage of *Taenia saginata*. Tropical Animal Health and Production. 8: 233−242, 1976.
7. Gallie GJ and Sewell MMH: Inoculation of calves and adult cattle with oncospheres of *Taenia saginata* and their resistance to reinfection. Tropical Animal Health and Production. 13: 147−154, 1981.
8. Gemmell MA, Matyas Z, Pawlowski Z and Soulsby EJL (eds): Guidelines on surveillance, prevention and control of Taeniasis/cysticercosis. World Health Organisation: VPH 83/49, 1981.
9. Gemmell MA: A critical approach to the concepts of control and eradication of echinococcosis/hydatidosis and taeniasis/cysticercosis. Proceedings of the Sixth International Congress of Parasitology, 469−472, 1986.
10. Gibbens JC, Harrison LJS and Parkhouse RME: Immunoglobulin class response to *Taenia taeniaeformis* in susceptible and resistant mice. Parasite Immunology. 8: 491−502, 1986.
11. Harrison LJS and Parkhouse RME: Antigens of taeniid cestodes in protection, diagnosis and escape. Current Topics in Microbiology and Immunology, 120: 159−168, 1985.
12. Harrison LJS and Parkhouse RME: Passive protection against *Taenia saginata* by a mouse monoclonal antibody reactive with the surface of the invasive oncosphere. Parasite Immunology, 8: 319-332, 1986.
13. Harrison LJS and Parkhouse RME: Identification of protective antigens in *Taenia saginata* cysticercosis. Immunobiology and Molecular Biology of Cestode Infections. Coopers Animal Health Symposium. Melbourne. A2, 1986.
14. Harrison LJS, Gallie GJ and Sewell MMH: Absorption of cysticerci in cattle after treatment of *Taenia saginata* cysticercosis with praziquantel. Research in Veterinary Science. 38: 378−380, 1984.

15. Harrison LJS, Parkhouse RME and Sewell MMH: Variation in 'target' antigens between appropriate and inappropriate hosts of *Taenia saginata* metacestodes. Parasitology. 88: 659–663, 1984.

16. Heath DD and Harrison GBL: Protective antigen sources from *Taenia taeniaeformis* larvae and from larval cestodes in sheep. Immunobiology and Molecular Biology of Cestode Infections. Coopers Animal Health Symposium. Melbourne. A6, 1986.

17. Joshua GWP, Harrison LJS and Sewell MMH: *Taenia saginata* metacestodes: immunochemical analysis. Transactions of the Royal Society of Tropical Medicine and Hygiene, (in press).

18. Lightowlers MW, Honey RD, Rickard MD and Mitchell GF: Development of a model vaccine against cysticercosis. Immunobiology and Molecular Biology of Cestode Infections. Coopers Animal Health Symposium. Melbourne. A4, 1986.

19. Lloyd S: Homologous and heterologous immunization against metacestodes of *Taenia saginata* and *Taenia taeniaeformis* in cattle and in mice. Zeitschrift fur Parasitenkunde. 60: 87–96, 179.

20. Lloyd S and Soulsby EJL: Passive transfer of immunity to neonatal calves against metacestodes of *Taenia saginata*. Veterinary Parasitology. 2: 355–362, 1976.

21. Parkhouse RME and Harrison LJS: Cyst fluid and surface associated antigens of *Taenia* sp. metacestodes. Parasite Immunology, (in press).

22. Rickard, MD and Brumley JL: Immunization of calves against *Taenia saginata* infection using antigens collected by *in vitro* incubation of *T. saginata* oncospheres or ultrasonic disintegration of *T. saginata* and *T. hydatigena* oncospheres. Research in Veterinary Science. 30: 99–103, 1981.

23. Soulsby EJL and Lloyd S: Passive immunization in cysticercosis: characterization of the antibodies concerned. Cysticercosis: present state of knowledge and current perspectives. Academic Press. 539–548, 1982.

24. Urquhart GM: Epizootiological and experimental studies on bovine cysticercosis in East Africa. Journal of Parasitology. 47: 857–869, 1961.

AN IMPORTANT FOCUS OF PORCINE AND HUMAN CYSTICERCOSIS IN WEST CAMEROON

A.ZOLI, S.GEERTS & T.VERVOORT

ABSTRACT

Since there were several indications that T.solium cysticercosis is an important zoonosis in the Menoua region (West Cameroon), a more detailed study was carried out to get a better idea of the prevalence of the parasite in pigs and men from this area. Ante mortem inspection of about 600 pigs on several local markets showed that 24.6% harboured cysticerci. By classical meat inspection of 151 pigs, taken at random at 5 different slaughterhouses in the region, 19.9% of the pigs were found positive. Using Elisa, however, antibodies against T.solium metacestodes were shown at least in 38% of these pigs.

A sero-epidemiological survey of 764 human serum samples, taken from apparently healthy people at different villages in the region, showed that 15.1% reacted positively in Elisa with a T.crassiceps metacestode antigen. After absorption of these sera to eliminate cross-reactions with filariasis and echinococcosis and comparative evaluation of the reactions with different other parasite antigens, al least 2.4% of the samples still remained positive for cysticercosis. These high levels of infection as well in men as in pigs prove that the Menoua region is an important focus of T.solium cysticercosis.

1. INTRODUCTION

Cysticercosis caused by the metacestodes of Taenia solium is a serious health and economic problem in some regions of Africa. Up to now, however, very few data are available about the distribution of T.solium cysticercosis in the latter continent. Nelson et al.(10) reported in 1965 that T.solium was common in Madagascar, the Cameroons, parts of East-Congo (Zaïre) and South Africa. Few research has been done since then. Pandey & Mbemba (11) confirmed in 1976 that 0.1-8.1% of the pigs in different regions of Zaïre were infected with Cysticercus cellulosae and there are records of human cysticercosis as well (12).

The disease in man was recorded for the first time in West Africa in Ivory Coast (1). Proctor et al.(13) identified spinal cysticercosis as a major cause of paraplegia in Ghana. In Tchad (7) and Nigeria (3, 4) high prevalence figures in pigs were mentionned in some regions, although no information is available on human cysticercosis. In Cameroon 2 short reports (8, 9) have been published on taeniasis-cysticercosis due to T.solium. Since there were several indications that T.solium cysticercosis is an important zoonosis in the Menoua division (West Cameroon) this study was undertaken to collect more detailed information on the prevalence of porcine and human cysticercosis in this area.

2. MATERIAL AND METHODS

2.1. Detection of T.solium cysticercosis in pigs.

2.1.1. Ante-mortem inspection was carried out from september 1984 until may 1985 on 607 pigs, selected at random on 6 livestock markets (Dschang-city, Bafou, Baleveng, Penka-michel, Fongo-Tongo and Fongo-Ndeng) and in 5 slaughterhouses (the same localities, except Penka-michel) of the Menoua division. The animals were examined for the presence of cysticerci by inspection of the inferior side of the tongue and of the conjunctiva of the eyes (18).

2.1.2. Post-mortem inspection of 151 out of the 607 abovementioned pigs was carried out by making incisions in the internal and external masseters, the heart, the tongue, the diaphragm , the muscles of the fore legs and hindlegs and also the psoas (18).

2.1.3. Serology. Two hundred serum samples were taken from pigs, kept by smallholders according to local customs (traditional husbandry). The blood was taken in the same abattoirs,which were mentioned above.
The sera were examined by the ELISA-test and compared with the sera of 32 pigs, originating from the farm of the University Centre of Dschang, where the pigs were kept inside throughout the year (according to modern hus-bandry practices), so that infection with cysticercosis was unlikely. A serum sample was considered as positive when the optical density (O.D.) exceeded the treshold value, which was calculated on the basis of the mean O.D. + 3 x the standard deviation (S.D.) of the pig sera from the university farm. The optimal dilutions and incubation times of antigen, serum and conjugate were determined by checkerboard titration. Delipidised hydrosoluble extracts of T.crassiceps and T.solium metacestodes were used as antigens. The latter were recovered from local naturally infected pigs. The antigens were prepared according to the method used by Geerts et al. (6), except that the cyst fluid of the T.solium metacestodes was not incor-porated in the antigen, as suggested by Flisser & Larralde (5).

The Elisa-test was performed essentially as described by Geerts et al. (6). Briefly the antigens (5 µg protein/ml) were incubated overnight at 4°C. Serum was diluted 1/200 and conjugate (R a SW IgG (H + L)/PO, Nordic) 1/500 in PBS-Tween 20 with addition of respectively 2 and 4% normal horse serum. Orthophenylene diamine was used as substrate.

2.2. Detection of T.solium cysticercosis in men.

A total of 764 serum samples were taken a) in the following villages of the Menoua-region: Baleveng; Bafou-road, -Chefferie, -Souschefferie, -Pastorale and -Cooperative; Bamendou II;

b) from the workers and students on the university farm (63) and

c) from patients (26) at the hospital of Dschang, who were not suspected for cysticercosis.
The majority of the examined people were adults (725 or 94.9%). Only 39 children were included. The number of samples taken from males and females was respectively 324 (42.4%) and 440 (57.6%).

Twenty serum samples from black Africans without parasitic infections were used as negative controls. These were used to calculate the treshold value (mean O.D. + 3 x S.D). The Elisa-test was carried out using serum diluted 1/200 and conjugate (GaH IgG (H + L)/PO, Miles), diluted 1/5000 in PBS-Tween 20. A fraction (F_1) of T.crassiceps metacestodes, prepared by gel filtration on Ultragel (AcA 34; LKB) was used as antigen (at a concentration of 5 µg protein per ml) in order to improve the specificity of the test.

All the positive sera were further tested against a battery of 7 different antigens: total extracts of Ascaris suum, Fasciola hepatica, Litomosoides carinii,Cysticercus cellulosae and Schistosoma mansoni; hydatid fluid of Echinococcus granulosus (horse) and excretion-secretion products of Toxocara canis. In order to differentiate between filariasis and echinococcosis absorption tests were performed essentially following the technique described by Speiser (14). For the confirmation of true cysticercosis or echinococcosis cases absorption of the sera was done with T.crassiceps metacestode or E.granulosus hydatid fluid antigens. Indirect haemagglutination was carried out with this latter antigen using a commercial kit (Fumouze, France). A titre of 1/512 was considered as positive.

3. RESULTS AND DISCUSSION
3.1. Porcine cysticercosis.

Tables 1 and 2 show the results of the ante-mortem and post-mortem inspection of the local pigs.

TABLE 1. Prevalence of T.solium cysticercosis in pigs of the Menoua division as detected by ante-mortem inspection.

	Total No. of pigs examined	Pigs infected with T.solium	
		number	%
Adult pigs ($>$ 1 year)	409	102	24.9
Young pigs ($<$ 1 year)	198	47	23.7
Total No. of pigs	607	149	24.6

TABLE 2. Prevalence of T.solium cysticercosis in pigs of the Menoua division as detected by post-mortem infection.

	Total No. of pigs examined	Pigs infected with T.solium	
		number	%
Adult pigs ($>$ 1 year)	142	27[+]	19
Young pigs ($<$ 1 year)	9	3[+]	33.3
Total No. of pigs	151	30	19.9

[+] 24 out of 27 adult pigs and all 3 young pigs were negative at ante-mortem inspection.

From the tables 1 and 2 it is clear that there is a high prevalence of pig cysticercosis in the Menoua region. The results of the meat inspection are comparable to the highest figures (18.4%) found in Nigeria (3, 4) and are even higher than those found in hyperendemic areas in Mexico (15). The fact that the figures from ante-mortem inspection are higher than those of post-mortem inspection is a common feature in many countries since the owners bring their pigs to the official abattoir only when no cysts are present in the mouth.

Table 3 summarises the serological results of 200 pig sera in Elisa.

TABLE 3. Presence of antibodies in pig sera against homologous and heterologous taeniid antigens (Elisa).

Metacestode antigen	No. of sera examined	Positive sera for cysticercosis	
		No.	%
T.crassiceps	200	106[+]	53
T.solium	200	89[+]	44.5

[+] 76 (38%) sera were positive for both antigens simultaneously

As expected these figures are still much higher than those from ante-or post-mortem inspection. Three reasons can be advanced to explain this: 1. there are surely a number of cross-reactions with other parasites, 2. some pig sera may contain antibodies, although no cysticerci are present (any more) and 3. a number of infected pigs are not detected by meat inspection (WHO 1983). It can be concluded that the figure of 24.6% of infected pigs, as evidenced by ante-mortem inspection is certainly a conservative figure and the real infection rate lies probably between 24.6 and 38%, the lowest figure of the serological results.

3.2. Human cysticercosis

Out of 764 serum samples 115 (15.1%) were shown to be positive in Elisa using fraction F_1 of a T.crassiceps metacestode antigen and a treshold level with a 99.7% confidence limit (mean O.D. + 3.SD = 0.286 + 3 x 0.127 = 0.667). The majority of these positive sera, however, were identified as cross-reactions after comparative evaluation against different parasite antigens and absorption tests with heterologous antigenic material (Table 4).

TABLE 4. Serodiagnosis of 115 human sera, positive[+] for
cysticercosis, after comparative evaluation against
different parasite antigens in Elisa.

	T.solium cysticerco-sis	Filariasis	Hydatidosis	Ascaridiasis and/or visceral larva migrans	Du-bious	Nega-tive
antigens	C.cellulo-sae	Litomosoides carinii	E.granulosus hydatid fluid	Ascaris suum Toxocara canis (E.S.)		
No. of sera	17°	76°	1[++]	3	10	8°

[+] positive in Elisa with T.crassiceps metacestode antigen

[++] confirmed by indirect haemagglutination test and absorption tests

° confirmed by absorption tests

Table 4 shows that at least 17 sera, i.e. 2.4% of the total number of
tested sera can be considered as true positives for cysticercosis. Since
it is known that at least 10 to 20% of known positive cases are not detected
by a sensitive technique like Elisa (5), the real frequency of human
cysticercosis in the region is certainly higher than 2.4%. Another reason
why the figure of 2.4% is probably an underestimation of the reality is
that only apparently healthy people have been sampled and not those showing
characteristic symptoms of neurocysticercosis. When this figure is compared
with the results from other sero-epidemiological studies it is very similar
to the figures from hyperendemic areas (Central America, some regions in
the Far East). In a large survey in Mexico only 0.45% of about 20.000 examined serum samples were found positive with immunoelectrophoresis, but it is
known that this technique is not very sensitive and detects only half of
the cases (16). In another serological survey in several areas of South
East Asia the percentages of positive reactions in Elisa varied between 3
and 21% according to the region (2). These latter figures,however,were not
corrected for cross-reactions.
    The results of this study confirm the observations, made by Marty et al.
(9) on 95 people living in the same region of Cameroon. These authors also
detected antibodies against T.solium metacestodes in 2% of the examined
population. In order to obtain a more complete picture of the situation
concerning human cysticercosis in the Menoua division, it is necessary,
however, to collect more data about the number of T.solium carriers among
the local population and also the number of people showing the typical
symptoms of cysticercosis (epilepsy, psychiatric symptoms etc.) Data about
the presence of cysticerci at routine autopsy would also be very interesting. Based on the currently available figures, however, it can be concluded
that the Menoua region is a very important focus of T.solium cysticercosis.
This can easily be explained since all the conditions are present for a
direct human-to-pig transmission of the parasite:

1. Open air fecalism or deliberate defecation in the pig sties are common practices in the region.
2. Well trained and qualified meat inspectors are lacking and clandestine slaughtering of pigs is common.
3. Consumption of raw or insufficiently cooked pork is common in the region.
4. Detection and treatment of human T.solium carriers is often lacking.

ACKNOWLEDGEMENTS
This work has been partly financed by a grant of the Institute for Scientific Research in Industries and Agriculture (IWONL, Belgium). The excellent technical assistance of F.Ceulemans and N.Aerts is gratefully acknowledged. The authors also wish to thank Dr.J.Brandt for his help in various ways.

REFERENCES
1. Bowesman C: Cysticercosis in West Africa. Annals of Tropical Medicine and Parasitology, 46: 101-102, 1952.
2. Coker-Vann MR, Subianto OB, Brown P, Diwan AR, Desowitz R, Garruto RM, Gibbs CJ and Gajdusek DC: Elisa antibodies to cysticerci of Taenia solium in human population in New Guinea, Oceania, and Southeast Asia. Southeast Asian Journal of Tropical Medicine and Public Health, 12: 499-505, 1981.
3. Dada BJO: Taeniasis, Cysticercosis and Echinococcosis/Hydatidosis in Nigeria: II. Prevalence of bovine and porcine cysticercosis and Hydatid disease in slaughtered food animals based on retrospective analysis of abattoir records. Journal of Helminthology, 54 (4): 287-292, 1980.
4. Dada BJO: Taeniasis, Cysticercosis and Echinococcosis/Hydatidosis in Nigeria: III. Prevalence of bovine and porcine cysticercosis and hydatid cyst infection based on joint examination of slaughtered food animals. Journal of Helminthology, 54 (4): 293-297, 1980.
5. Flisser A & Larralde: in "Immunodiagnosis of parasitic diseases. Vol.I. Helminthic diseases".Eds. Walls and Schantz p. 109-162, 1986.
6. Geerts S, Kumar V, Ceulemans F & Mortelmans J: Serodiagnosis of Taenia saginata cysticercosis in experimentally and naturally infected cattle by enzyme linked immunosorbent assay. Research in Veterinary Science,30: 288-293, 1981.
7. Graber M & Chailloux A: Existance au Tchad de la ladrerie porcine à Cysticercus cellulosae (Rudolphi). Revue d'Elevage et de Médecine Vétérinaire des pays tropicaux, 23: 49-55,1970.
8. Marty P, Herzog U, Marty-Jaussan I, Le Fichoux Y et Doucet J: Deux cas de cysticercose observés au Cameroun. Médecine Tropicale, 45: 83-86, 1985
9. Marty P, Mary C, Pagliardini G, Quilici M et Le Fichoux Y: Courte enquête sur la cysticercose et le taeniasis à Taenia solium dans un village de l'Ouest Cameroun. Médicine Tropicale, 46: 181-185, 1986.
10. Nelson GS, Pester FR and Rickman R: The significance of wild animals in the transmission of cestodes of medical importance in Kenya. Transactions of the Royal Society of Tropical Medicine and Hygiene, 59: 507- ,165.
11. Pandey VS et Mbemba Z: Cysticercosis of pigs in the Republic of Zaïre and its relation to human taeniasis. Annales de la Société belge de Médecine Tropicale, 51: 43-46, 1976.

12. Pieters G: Cysticercosis in Africans in the Belgian Congo- 7 cases seen in Madembu Region. Annales de la Société Belge de Médecine Tropicale, 35: 751-755, 1955.

13. Proctor EM, Powell SJ & Elsdon-Dew R: The serological diagnosis of cysticercosis. Annals of Tropical Medicine and Parasitology,60:146,1966.

14. Speiser F: Application of the enzyme-linked immunosorbent assay (Elisa) for the diagnosis of filariasis and echinococcosis. Tropenmedizin und Parasitologie, 31: 459-466, 1980.

15. Velasco-Suarez M, Bravo-Becherelle MA & Quirasco F: Human cysticercosis: medical-social implications and economic impact. in "Cysticercosis. Present state of knowledge and perspectives" Eds.Flisser A et al. p.47-52, 1982.

16. Woodhouse E, Flisser A and Larralde C: Seroepidemiology of human cysticercosis in Mexiko. in "Cysticercosis. Present state of knowledge and perspectives". Eds Flisser A et al. p.11-24, 1982.

17. W.H.O.: Guidelines for surveillance, prevention and control of taeniasis/cysticercosis. Eds.Gemmell M, Matyas Z, Pawlowski Z & Soulsby EJL. p.78, 1983.

18. Zoli A: Prevalence de la cysticercose porcine dans le departement de la Ménoua. Annales du Centre Universitaire de Dschang (soumis).

CYSTICERCUS FASCIOLARIS IN MICE: A LABORATORY MODEL FOR SELECTING NEW
DRUGS ON CYSTICERCOSIS

O.VANPARIJS, L.HERMANS and H.LAUWERS

1. ABSTRACT

In this study we describe the use of Cysticercus fasciolaris, the
larval stage of the cat tapeworm Hydatigera taeniaeformis as model for
testing anthelmintics in mice. Albino Swiss mice were artificially
infected by gavage with eggs of H.taeniaeformis and medicated with
mebendazole or flubendazole in a prophylactic or therapeutic test.
In the prophylactic test, mice were treated in groups of 10 with both
substances at 250, 125, 63 and 32 ppm from day -2 to day 21. In the
therapeutic test the medication started on day 21 through day 42 at dose
levels of 250, 125 and 63 ppm. The autopsy was performed 3 weeks after
the end of the treatment. Mebendazole prevented cyst development at dose
levels of 63 ppm and higher. The 32 ppm dose had little or no
prophylactic activity. In the therapeutic test a dose of 125 ppm was
needed to kill all cysts. In the 63 ppm group 4 mice were still positive
but the number of living cysts was highly reduced compared to the control
mice. Flubendazole was prophylactically 100% active at 63 ppm and higher.
At 32 ppm the cyst growth was completely inhibited in 4 mice and highly
reduced in 6 mice. The therapeutic activity was complete at dose levels
of 250 and 125 ppm and 1 mouse out of 10 treated at 63 ppm had only 1
living cyst.
Furthermore, the results in this laboratory study obtained with both
benzimidazoles were compared to those obtained by many workers in large
animals, especially in pigs, infected with Cysticercus cellulosae or
Cysticercus tenuicollis. Flubendazole was 100% active in pigs infected
with C.cellulosae after 14 days treatment at least 125 ppm in the food.
Mebendazole was found to be highly efficient against C.tenuicollis in
pigs treated at 25 mg/kg B.W., administered in feed for 5 days.

2. INTRODUCTION

A wide variety of methods are used to screen anthelmintics in
laboratory animals. For nematodes, the models generally used, are
Nippostrongylus braziliensis and Strongyloides ratti in rats and
Nematospiroides dubius in mice. For trematodes, a strain of Schistosoma
mansoni in mice can be used to screen drugs on bilharziosis and Fasciola
hepatica in rats or mice for screening on distomatosis. For cestodes, the
models, mostly used, are Hymenolepis nana in mice and Hymenolepis
diminuta in rats.
Systemic activity of new drugs can be demonstrated in models using larval
stages of Trichinella spiralis in rats or mice and larval stages of
Ascaris suum in mice.

For activity on larval stages of cestodes, a large scale of models can be used. Metacestodes of Taenia crassiceps can be used in mice, Echinococcus granulosus cysts in mice, Echinococcus multilocularis metacestodes in meriones, Tetrathyridium larvae in mice, Cysticercus pisiformis in rabbits and Cysticercus fasciolaris in rats and mice.

The aim of this study is to describe the use of C.fasciolaris in mice as a model for testing the activity of mebendazole and flubendazole. At the same time information is given on the activity of mebendazole and flubendazole against cysticercosis in pigs.

3. MATERIAL AND METHODS

3.1. Animals

Albino Swiss mice, 20 days old, were used in this study. They were divided in groups of 10 mice. Feed pellets and water were available ad libitum.

3.2. Infection

The mice were infected orally by gavage with 100 Hydatigera taeniaeformis eggs, collected from gravid proglottids expelled by artificially infected cats. The eggs were prepared after maceration, washing in saline and sieving through a 200-mesh sieve (75 mu).

3.3. Medication

Both benzimidazoles were administered through a pelleted commercial mouse diet at dosages of 250, 125, 63 and 32 ppm. The animals were fed the medicated pellets ad libitum either from day -2 to day 21 after infection for the prophylactic activity or from day 21 to day 42 for the therapeutic activity. Control mice were fed the unmedicated feed.

3.4. Autopsy

All animals were autopsied 3 weeks after the end of the medication period. The liver was removed for counting the normally developed cysts, dead cysts or granulomas of C.fasciolaris.

4. RESULTS

The results of medication are summarized in the Tables 1 to 4. As shown in Table 1, mebendazole at 63 ppm and higher prevented cyst development in all mice. A 32 ppm dose had little or no prophylactic activity.

Table 1: Prophylactic activity of mebendazole on *Cysticercus fasciolaris* in mice. Medication from day-2 to day 21 (10 mice per group).

Dose ppm	Number of mice with cysticerci	granuloma	Mean number (limits) of cysticerci	granuloma
250	0	2	0	0.4 (0- 3)
125	0	5	0	1.1 (0- 6)
63	0	8	0	2.1 (0- 6)
32	8	4	13.8 (0-28)	1.5 (0- 3)
0	10	6	19.1 (2-35)	10.8 (0-34)

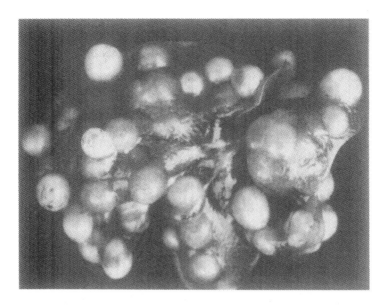

Figure 1. Cysticercus fasciolaris in murine liver

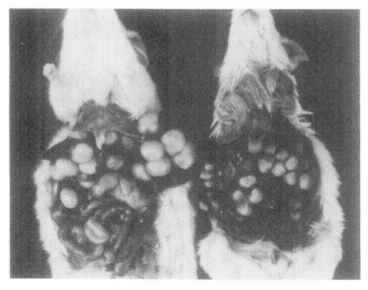

Figure 2. Cysticercus fasciolaris in control mouse (left) and in mouse treated with mebendazole (right)

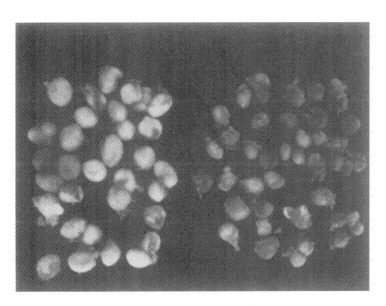

Figure 3. Isolated <u>Cysticercus fasciolaris</u> cysts from untreated (left) and treated mouse (right)

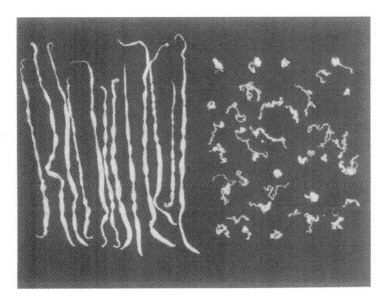

Figure 4. Evaginated <u>Cysticercus fasciolaris</u> from untreated mouse (left) and dead necrotic cyst material from treated mouse (right)

The therapeutic effect of mebendazole, as shown in Table 2, was manifested at 125 and 250 ppm, since no living cysts were observed at these dose levels. In 4 mice with living cysts in the 63 ppm group, the mean number of cysts was highly reduced as compared to the controls (1.9 to 18.3).

Table 2: Therapeutic activity of mebendazole on *Cysticercus fasciolaris* in mice. Medication from day 21 to day 42 (10 mice per group).

Dose ppm	Number of mice with living cyst.	Number of mice with dead cyst.	Mean number (limits) of living cyst.	Mean number (limits) of dead cyst.
250	0	10	0	15.0 (7-21)
125	0	10	0	19.4 (7-36)
63	4	10	1.9 (0- 8)	16.2 (3-33)
0	10	3	18.3 (2-43)	1.8 (0-14)

As for mebendazole, the prophylactic efficacy of flubendazole, as shown in Table 3, was 100% at dose levels of 63 ppm and higher. At 32 ppm cyst growth was completely inhibited in 4 mice and highly reduced in 6 mice.

Table 3: Prophylactic activity of flubendazole on *Cysticercus fasciolaris* in mice. Medication from day-2 to day 21 (10 mice per group).

Dose ppm	Number of mice with cysticerci	Number of mice with granuloma	Mean number (limits) of cysticerci	Mean number (limits) of granuloma
250	0	0	0	0
125	0	5	0	1.4 (0- 4)
63	0	3	0	0.6 (0- 3)
32	6	5	3.4 (0-12)	1.8 (0- 7)
0	9	8	12.9 (0-33)	10.2 (0-25)

As summarized in Table 4, the therapeutic activity of flubendazole was complete at dose levels of 125 and 250 ppm. In one out of 10 mice medicated at 63 ppm only 2 living cysts could be recovered.

Table 4: Therapeutic activity of flubendazole on *Cysticercus fasciolaris* in mice. Medication from day 21 to day 42 (10 mice per group).

Dose ppm	Number of mice with		Mean number (limits) of	
	living cyst.	dead cyst.	living cyst.	dead cyst.
250	0	10	0	16.5 (5-35)
125	0	10	0	12.2 (3-23)
63	1	10	0.2 (0- 2)	17.0 (5-28)
0	10	2	17.9 (4-36)	1.2 (0- 8)

In the 4 experiments all but one control mice were moderately to heavily infected with living cysticerci. The mean number varied from 12.9 to 19.1 cysts per mice.
The average daily feed consumption in the 8 medicated groups of the prophylactic treatment was 4.9 g per mouse. Based on a mean body weight of 23 g, the daily drug intake was 52 mg/kg (250 ppm), 26 mg/kg (125 ppm) and 13 mg/kg (63 ppm) respectively. In the therapeutic treatment the average daily feed consumption of all medicated groups was 6.5 g per mouse and the mean body weight was 30 g. Based on this body weight the daily drug intake was 54 mg/kg (250 ppm), 27 mg/kg (125 ppm) and 23.5 mg/kg (63 ppm) respectively.

5. DISCUSSION
    The present findings show that both benzimidazoles have a lethal effect on C.fasciolaris up to doses of 63 ppm. Flubendazole was found to be more active than mebendazole since the lowest prophylactic dose of 32 ppm flubendazole showed a mean number of 3.4 living cysticerci in 6 mice compared to 13.8 cysticerci in 8 mice treated with mebendazole. Moreover, the therapeutic dose of 63 ppm flubendazole cured 9 out of 10 mice compared to 6 out of 10 animals treated with mebendazole.
Other authors have reported the effect of mebendazole and flubendazole on larval stages of E.granulosus (7,10) and E.alveolaris (2,3) in mice.
The excellent results with both benzimidazoles in these laboratory models have been confirmed by many workers in large animals.Tellez-Giron et al.(9) reported in 1981 the activity of flubendazole on Cysticercus cellulosae in 15 naturally infected pigs. Oral doses of 125 and 500 mg, administered in gelatine capsules twice a day for 10 days, were lethal to all the cysticerci. The viability of the cysts was tested in vitro. In another Mexican study Galdamez Toledo (4) showed that flubendazole was found to be 100% effective against C.cellulosae in 6 pigs medicated in the diet at doses of 125 and 500 mg for 14 consecutive days.

The larvicidal activity of flubendazole has also been reported by
Guerrero et al.(5) in a study on 18 pigs naturally infected with
C.cellulosae and medicated at 30, 125 and 500 ppm for 14 consecutive
days. The treatment resulted in a high variability in the number of
viable cysts but 5 out of 8 pigs, treated ad libitum at 500 ppm, showed
all cysticerci dead and non-viable.
Mebendazole has been used successfully in humans to combat larval stages
of E.granulosis (1) and E.alveolaris (8). Hörchner et al.(6) belong to
the few authors reporting the efficacy of mebendazole in pigs. He studied
the therapeutic activity in 21 pigs artificially infected with
Cysticercus tenuicollis and obtained excellent results with a dose of 25
mg/kg, administered in a daily portion of feed for 5 days. At 5 mg/kg
given for 14 days, only 3 out of 7 pigs were found to be negative
although the reduction of living cysticerci was highly significant as
compared to the unmedicated controls.

6. CONCLUSION

From tnese data it is clear that C.fasciolaris in mice can be used as
a laboratory model for possible screening of compounds by dietary
administration. Infective eggs of H.taeniaeformis can easily be collected
from infected cats which can produce eggs for several months. Mice are
highly susceptible for C.fasciolaris, since in this study 39 out of 40
control mice were positive. Infective mice can be used for several weeks
and the morbidity is nihil.

7. ACKNOWLEDGEMENTS

This investigation was supported in part by the "Instituut tot
Aanmoediging van het Wetenschappelijk Onderzoek in Nijverheid en
Landbouw" (IWONL), Brussels. The authors thank J.Frederickx for preparing
and Y.Devocht for typing the manuscript.

8. REFERENCES

1. Bekhti A, Schaaps JP, Capron M, Dessaint SP, Santora F and Capron A:
   Treatment of hepatic hydatid disease with mebendazole. Preliminary
   results in four cases. British Medical Journal 2: 1047-1051, 1977.
2. Campbell WC, McCracken RO and Blair LS: Effect of parenterally
   injected benzimidazole compounds on Echinococcus multilocularis and
   Taenia crassiceps metacestodes in laboratory animals. Journal of
   Parasitology 61: 844-852, 1975.
3. Eckert J and Pohlenz J: Zur Wirkung von Mebendazol auf Metazestoden
   von Mesocestoides corti und Echinococcus multilocularis.
   Tropenmedizin und Parasitologie 27: 247-262, 1976.
4. Galdamez Toledo OE: Ensayo preliminar del flubendazol contra
   Cysticercus cellulosae en cerdos. Thesis, Universidad Nacional
   Autonoma de México, 1980.
5. Guerrero J, Naranjo FG and Chiquillo de Reinoso ME: Study of the
   anti-Cysticercus cellulosae activity of flubendazole in swine.
   Pitman-Moore U.S.A., Protocol No.FLU-24, February 1982.
6. Hörchner F, Langnes A and Oguz T: Die Wirkung von Mebendazole und
   Praziquantel auf larvale Taenienstadien bei Maus, Kaninchen und
   Schwein. Tropenmedizin und Parasitologie 28: 44-50, 1977.
7. Kammerer WS and Judge DM: Chemotherapy of hydatid disease
   (Echinococcus granulosus) in mice with mebendazole and bithionol.
   American Journal of Tropical Medicine and Hygiene 25: 714-717, 1976.

8. Ludin CE, Gyr K and Karoussos K: Therapy of alveococcosis in man. Journal of International Medical Research 5: 367-368, 1977.
9. Tellez-Giron E, Ramos MC and Montante M: Effect of flubendazole on Cysticercus cellulosae in pigs. American Journal of Tropical Medicine and Hygiene 30: 135-138, 1981.
10. Vanparijs O: Chemotherapy of experimental echinococcosis in mice. Annals of Tropical Medicine and Parasitology 80: in press, 1986.

LARGE-SCALE USE OF CHEMOTHERAPY OF TAENIASIS AS A CONTROL MEASURE
FOR TAENIA SOLIUM INFECTIONS

Z.S PAWLOWSKI

1.   INTRODUCTION

Today's epidemic situation in Taenia solium taeniasis/cysticercosis
in some areas of Latin America, South Africa and South-East Asia resembles
that in Germany at the end of the 19th century (Nieto, 1982). For example,
in 18 Latin American countries the overall prevalence of human neurocysticercosis based on post mortem examinations was estimated to be 0.1%
(Schenone et al., 1982), but in some areas it exceeded 2%. In Mexico, the
prevalence rates for T. solium taeniasis varied between 0 and 7.1% depending on the locality; in some slaughterhouses up to 10% of the pigs
inspected had cysticercosis (Aluja, 1982). In Berlin in the 1860's
T. solium cysticercosis was found in 2% of the corpses examined post mortem
and in 1883 cysticercosis was detected in 0.6% of the pigs inspected after
slaughter.

At the present time in T. solium endemic areas preventive and control
measures are either not practised at all or follow the traditional style
focussing on infected pigs (Pan American Health Organization, 1980).
Recent changes in health policies, improved ways of communication, growing
experience in community-oriented chemotherapy with safe and effective drugs
all suggest that the emphasis in control measures should be focussed on
T. solium taeniasis in man as the major source of infection both for man
himself and for pigs.

2.   PAST EXPERIENCE IN TAENIASIS/CYSTICERCOSIS CONTROL MEASURES

It is unlikely that meat inspection alone was responsible for the
complete eradication of T. solium infections in Germany and most of Europe.
Much of the credit must be due to the changes in socioeconomic conditions,
mainly higher levels of personal hygiene and general sanitation as well as
changes from individual outdoor pig-rearing to indoor pig-farming
(Pawlowski, 1982). An important factor was also the lower consumption of
raw pork products because of the common fear of trichinellosis, which
caused several severe epidemics in Middle Europe in the late 1980's.

In Germany the process of T. solium eradication proceeded slowly;
there were still 0.03% of pigs infected in the year 1902 and 0.002 to
0.008% in the years 1925-1930. The slow progress in controlling taeniasis/
cysticercosis was mainly due to the persistence of a reservoir of T. solium
infections in man; new infections were to a great extent prevented by meat
inspection but most of the already existing infections in man remained

untreated. Up to the 1950's the chemotherapy of human taeniasis, based on natural products (e.g. Koso flowers, Areca nuts, pumpkin seeds and male fern extracts), was toxic, ineffective and even unsafe.

3. IMPROVEMENTS IN CONTROL TOOLS

The major breakthrough in chemotherapy of human taeniasis came with the use of tin compounds and mepacrine in the 1950's (Pawlowski, 1970), with the introduction of niclosamide in 1960 (Pearson and Hewlett, 1985) and praziquantel in the late 1970's (Davis and Wegner, 1979; Groll, 1980). The drugs of choice for taeniasis today are niclosamide and praziquantel. Niclosamide has been used so far in more than 10 million patients without any cases of severe side-effects being reported in a post-marketing survey; however, its efficacy was always dependent on the quality of the drug. Praziquantel meets all the modern drug safety requirements and has been widely used in community-oriented schistosomiasis control projects.

Both niclosamide and praziquantel are safe, single-dose drugs, with an efficacy of over 90% and with a limited number of general contraindications, all of which makes mass-chemotherapy feasible.

Other major improvements in taeniasis/cysticercosis control tools within the last 100 years are the developments in communication systems and the changes in health policies. Radio, television, video, a reduced rate of analphabetism, the increased role of women in human contacts – all these factors help to improve the effectiveness of health education. Integrated health and general development programmes, trends towards equity in health services distribution and availability, the primary health care approach, and emphasis on preventive medicine, all give solid support to community-oriented control programmes.

Modern approaches to taeniasis/cysticercosis control measures have been elaborated (World Health Organization, 1983). Very recent experience in the large-scale use of chemotherapy may give rise to some improvements in the general design of control measures.

4. PRESENT SITUATION IN T. SOLIUM ENDEMIC AREAS

Unfortunately, none of the control measures which have previously contributed to the eradication of T. solium infections in most of Europe, such as improved general economic and sanitary conditions, regular meat inspection and indoor pig-husbandry, can be promptly and effectively introduced today into taeniasis/cysticercosis endemic areas. For example, in Latin America meat inspection covers only 15% of the pigs raised in small farms (Acha and Aguilar, 1964). There are still some economic factors against meat inspection (a pig found to be infected has to be sold at a lower price and a carcass found to be infected has to be destroyed). In Mexico, one-third of the pigs in rural areas have access to human faeces (Aluja, 1982); in some areas of Loja Province, Ecuador, even more than half the pigs are raised in insanitary conditions. With an increased demand for pork, raising pigs is an important source of income for individual farmers; keeping pigs roaming outdoors substantially reduces the

cost and work needed for their feeding, watering and cleaning. Industria-
lized pig-farming is rare in developing countries.

In view of all these circumstances, the traditional control measures,
based on general sanitation, zoohygiene and meat inspection and on education for health, will work only very slowly unless they are supported by a
wider use of chemotherapy against human taeniasis; only chemotherapy is
able to reduce, in a short time, the exposure of people and pigs to
T. solium cysticercosis.

In other words, the traditional control measures may remain as long-
term control projects aiming at the eradication of T. solium infections,
but the large-scale use of chemotherapy should be accepted as an essential
supportive measure, which will immediately reduce the exposure to human
cysticercosis, contamination of the environment and transmission of the
infection to pigs.

5.   USE OF LARGE-SCALE CHEMOTHERAPEUTIC MEASURES

Any large-scale chemotherapeutic intervention has to be based on the
results of a survey defining the local epidemiological situation. Surveys
can be based on records from diagnostic laboratories, hospitals and clinics
of post mortem examination and serological or coproscopical screening of
the human population as well as on in vivo or post-slaughter examination
of pigs. The survey methodology and diagnostic tools remain to be developed further, as they are still inadequate for the control needs.

In T. solium infections, there are three major epidemiological
patterns: (i) endemic foci of pig cysticercosis, human taeniasis and cysticercosis mainly in rural and periurban pig-breeding areas; (ii) foci of
human teaniasis and human cysticercosis without actual contact with the
foci in pigs, mainly in urban populations and immigrants; (iii) sporadic
cases of human cysticercosis in travellers, immigrants, consumers of
imported or locally contaminated food, etc., which cannot be traced back
to any known focus of taeniasis. Large-scale chemotherapy can be used only
in the first situation.

The endemic foci in man and pigs are usually concentrated in a few
villages (parts of the urban agglomerations as well) and/or in some
individual farms and families. Therefore, the large-scale chemotherapy
should be a village-oriented or family-oriented intervention. Mass chemotherapy of the whole village is justified in highly endemic areas which
can be identified by a high incidence of cysticercosis in man according to
hospital records or the results of a seroepidemiological survey, i.e.
T. solium taeniasis diagnosed in over 2% of the population and/or cysticercosis found in over 5% of the local pigs.

The results of community-oriented chemotherapy can be evaluated by
decreasing rates of cysticercosis in local pigs within a year from treatment. Then, depending on the outcome of this survey, the mass treatment
can be repeated or chemotherapeutic interventions oriented towards infected
farms or families. Such pilot, operational research has been organized by

the Parasitic Diseases Programme in Ecuador and in Mexico.

In endemic areas where meat inspection is available and animals marked, the infected pigs can be traced back to their original farm and chemotherapy should then be given to all the people living on the farm. A similar procedure should be carried out in all cases of taeniasis in man which has been diagnosed in the clinics or by coproscopical population screening. This technique was used to control T. saginata infections in Poland (Pawlowski, 1980) and has been suggested for Ecuador and Mexico in a later stage of the control projects.

In less endemic rural areas and in urban areas where no contacts can be established between human cases of taeniasis and pig cysticercosis, anthelmintic chemotherapy should be easily and perhaps freely available to anyone infected with either T. solium or T. saginata. Such action must be preceded by intensive health education campaigns and the mobilization of health services. Some countries have already introduced both obligatory and free treatment of human taeniasis (Pawlowski, 1980).

Wider use of chemotherapy against human taeniasis is often restrained by the lack of an effective taeniacide. It is frequently not understood that the amount of the drug needed for the treatment of one patient with neuro-cysticercosis would be sufficient to eradicate infection in 140 carriers of T. solium.

6.   EXPERIENCE FROM THE ECUADORIAN PROJECT

A project on the control of human neurocysticercosis by community-oriented treatment of human taeniasis in the Provinces of Loja and El Oro, Ecuador, was designed and funded by the Parasitic Diseases Programme of the World Health Organization in Geneva and carried out by the Centro de Investigacion y Entratamiento en Neurociencias (CIEN), Quito (principal investigator: Dr Marcelo Cruz), with the full cooperation of the national and local authorities.

These areas were selected for intervention because human taeniasis was frequently found in patients coming from the Province of Loja; over 1% of taeniasis has been reported from patients hospitalized in Loja Provincial Hospital and more than 5% of pig cysticercosis has been reported from the slaughterhouse in Loja. During the studies, 1117 pigs were examined in vivo: 3% of the pigs in the Province of Loja and 0.9% of those in the Province of El Oro were found to be infected. Out of 70 pigs slaughtered and inspected in Gonzanama, Loja Province, 11.4% were infected with cysticerci.

During the months of February and March 1986, 2446 households were visited and interviewed in the study area. It has been found that in many villages more than half the pigs had access to human faeces. At the same time, 3750 people were interviewed for neurological disorders; the prevalence of epilepsy was 12.4 per 1000, which is two to three times higher than the prevalences reported from the developed countries.

Of the total number of 13,290 people examined, 9,326 were treated
with a single dose of praziquantel given orally in a dose ranging between
3.4 and 8.7 mg/kg body weight (mean: 5.2 mg/kg). 149 people, i.e. 1.6% of
those treated had expelled Taenia. The number of people expelling Taenia
was probably higher in reality because some areas were not re-examined
after treatment, and also some people probably didn't wish to report that
they had expelled the parasite or some might have digested the tapeworm
without expelling proglottids. Some families had up to three persons
infected. The infection rates in the localities varied from 0% to 20.5%
of the local population infected. The treatment was well accepted and only
a few transient side-effects were reported.

In addition to the fact that the reservoir of Taenia infection in man
was reduced by at least 149 tapeworms in the study area, the chemotherapeutic intervention had three other important impacts: (i) it promoted
public interest in the problem of taeniasis/cysticercosis;(ii) it stimulated
local authorities' activities for the control of taeniasis/cysticercosis:
both provincial and local control committees have been appointed; (iii) it
promoted collection of data on sanitary and living conditions in the study
area, which can be used for long-term sanitation projects (WHO/CIEN
project 1986).

7.   CONCLUSIONS

The large-scale use of chemotherapy based on surveys,accepted by the
communities, coordinated with the promotion of sanitation, meat inspection
and health education, and backed by the authorities, will probably be a new
breakthrough in taeniasis/cysticercosis control measures in the areas where
T. solium infections are prevalent and human neurocysticercosis is a
serious health problem.

REFERENCES

1.   Acha PN and Aguilar FJ: Studies on cysticercosis in Central America
     and Panama. Am.J.Trop.Med.Hyg. 1964, 13:48-53.
2.   Aluja AS: Frequency of porcine cysticercosis in Mexico. In:
     Flisser A et al.(Eds) Cysticercosis: Present State of Knowledge and
     Perspectives. Academic Press, New York, 1982, 53-62.
3.   Davis A and Wegner DHG: Multicentre trials of praziquantel in human
     schistosomiasis: design and techniques. Bull.WHO 1979, 57: 767-771.
4.   Groll E: Praziquantel for cestode infections in man. Acta Trop.
     1980, 37: 293-296.
5.   Nieto D: Historical notes on cysticercosis. In: Flisser A et al.
     (Eds) Cysticercosis: Present State of Knowledge and Perspectives.
     Academic Press, New York, 1982, 1-7.
6.   Pan American Health Organization, Washington: Diagnosis of animal
     health in the Americas. Scientific Publication Nr 452, 1983, p.99.
7.   Pawlowski Z: Inorganic tin compounds as the alternative drug in
     human taeniarhynchosis. J.Parasitol. 1970, 56(Sect.2): 261.

8.   Pawlowski Z: (Studies on epidemiology of _Taenia saginata_ taeniasis
     and cysticercosis) In Polish. Wiadomosci Parazytologiczne 1980,
     26:539-552.
9.   Pearson RD and Hewlett EL: Niclosamide therapy for tapeworm
     infections. Ann.Intern.Med. 1985, 102(4):550-557.
10. Schenone H et al.: Epidemiology of human cysticercosis in Latin
     America. In: Flisser A et al. (Eds) Cysticercosis: Present State of
     Knowledge and Perspectives. Academic Press, New York 1982, 25-38.
11. World Health Organization: Guidelines for surveillance, prevention
     and control of taeniasis/cysticercosis, 1983, internal document
     VPH/83.49.
12. WHO/CIEN Programme on Control of Taeniasis/Cysticercosis.
     (Preliminary report), September 1986.

ZOONOTIC TREMATODIASIS IN SOUTH-EAST AND FAR-EAST ASIAN COUNTRIES

V. KUMAR

ABSTRACT

Zoonotic trematodiasis constitute a public health problem of considerable magnitude in many countries of South-east and Far-east Asia. These infections are caused by the digenetic trematodes of various genera and have a high rate of prevalence in the countries of their endemicity.

The trematode infections of the SE and FE Asian countries which are zoonoses and which involve the extra-intestinal organs of the mammalian host cause well known diseases as these affect the vital organs and cause spectacular pathology. The diseases like clonorchiasis, opisthorchiasis, paragonimiasis and schistosomiasis affecting the biliary system, lungs and hepato-intestinal tissues of the host need special mention. Clonorchiasis, opisthorchiasis and paragonimiasis are sea-food borne infections acquired through the ingestion of infected fishes, crabs and cray-fishes. Certain food habits of the ethnic communities have important bearing on their transmission dynamics. Schistosomiasis is water-borne and exposure of the paddy field farmers, bathing and doing laundry in waters contaminated with schistosome cercariae constitute major means of human infection.

The trematodes inhabiting the intestinal tract may also arouse varying grades of pathological effects. Gastrodiscoides hominis, Fasciolopsis buski and Artyfechinostomum malayanum are found in the intestine of humans and in all these three instances the local pigs of SE and FE Asia serve as the reservoir host. Close association of the farming communities with their local pigs assists their dissemination through the molluscan vector. Dogs and cats serve as hosts of certain zoonotic trematode infections like Heterophyes heterophyes, Metagonimus yokogawai, Haplorchis spp. etc. These intestinal trematode parasites are transmitted through ingestion of infected fishes. Echinostoma ilocanum, E.lindoense and E.hortense are other trematodes found in the intestine of humans.

Since the advent of praziquantel, the chemotherapy of many of the aforementioned infections may have been resolved to a great extent though other integrated control measures are indispensable in achieving a substancial reduction in their incidence.

1. INTRODUCTION

The infections and diseases which are caused by the digenetic trematodes and which are transmissible from vertebrate animal hosts to man and viceversa are endemic in certain geographical areas of the world. In the Southeast (SE) and Far-east (FE) Asian context, the zoonotic trematodiasis are well known not only because of the prevalence or the severity of the infection but also because the involved causative agents belong to diverse genera of parasitic trematodes. Fortunately or unfortunately, it is only

in the SE and FE Asian countries that zoonotic trematodiasis constitute
a public health problem of considerable magnitude and importance.

The food habit of a given population has important bearing on the
transmission dynamics of a number of food transmitted trematodiasis of
man (2, 29). This situation is best exemplified by the traditional food
habits of ethnic communities of SE and FE Asian countries. The sea-food
constitute important sources of dietary protein in human nutrition in
many countries of this area. Since the crabs, cray-fishes, snails including the fishes, which harbour infective metacercariae of various trematodes, are consumed raw or uncooked, this practice helps in dissemination
of the trematode infections. Again, in the absence of adequate sanitary
facilities, a large proportion of the rural population defecate in open
fields. There is indiscriminate disposal of faecal excreta which is used
as fertilizer or discharged into the streams where the snails are found.
Should these defecates contain eggs of trematodes, these again constitute
an important source of infection of various molluscan species which are
the obligatory first intermediate-hosts in the life-cycle of zoonotic
trematode parasites.

For convenience of description of the zoonotic trematodiasis of SE and
FE Asian countries this disease entity may be treated in two heads: namely,
extra-intestinal trematodiasis and intestinal trematodiasis. The following
account on the important zoonotic trematodiasis of SE and FE Asian countries is an attempt to outline some important features of these infections
and diseases though it is by no means complete.

2. EXTRA-INTESTINAL ZOONOTIC TREMATODIASIS

The extra-intestinal zoonotic trematodiasis of SE and FE Asian countries, viz., clonorchiasis,opisthorchiasis, paragonimiasis and schistosomiasis involve the vital organs of the host and, more often than not,
arouse considerable pathology. These affect the hepatic biliary system in
cases of clonorchiasis and opisthorchiasis,the pulmonary parenchyma in
case of paragonimiasis and the hepato-intestinal tissues in case of schistosomiasis.

2.1. Clonorchis sinensis infection (Clonorchiasis)

Clonorchiasis occurs in several countries of the FE Asian region and is
found in China, Japan, Korea, Taiwan and northern Vietnam. Certain areas
in southern China especially Kwangtung Province have high rates of human
infection. Heavy infections are found in and around Canton and Chaochow.
Certain fishes found in this region, which serve as second intermediate-
host and harbour the infective metacercaria, are principal food of the
local rural population and are also supplied to neighbouring areas and
Hong Kong. The rate of infection in humans in this region of China appears
to have declined because of improvement in sanitary conditions. Human
clonorchiasis in northern China was earlier thought to be absent but the
infection was detected in certain areas of this region in the late sixties.
In Japan, endemic areas of clonorchiasis have been known for a long time
around Kojima Bay (Okayama Prefecture) and the basin of the River Kitamani
(Miyagi Prefecture). Light infections with the trematode are reported near
the basin of the River Tone, Lake Kasumigaura, Nobi Plain in Aichi and Gify
Prefecture, Lake Biwa in Shiga Prefecture and the basin of the River Onga
and Chikugo. In Japan, in many endemic areas persons affected with clonorchiasis rarely display overt clinical effects. Clonorchiasis is highly

endemic in the Naktong, Kum, Nam, Sumgin and Hun river valleys in south
Korea and in Vietnam the disease is mainly present in the delta of the
Red River. In Taiwan, the areas around the Sun Moon lake and the county
of Kaohsiung are highly endemic for clonorchiasis (15).

The natural mammalian hosts of C. sinensis other than man include dogs,
cats, pigs, rats and camels (27). It appears that various fish eating
mammals can be infected experimentally though cats and dogs are important
reservoir hosts. In northern and central China the cats and dogs are frequently harbouring C. sinensis and the infection in human is relatively
rare. In Korea, however, dogs, pigs and rats are important reservoir
hosts and infection in cats is rare (14).

The operculate snail, Parafossarulus manchouricus, is the main molluscan
host of C. sinensis. It is widely distributed in Japan, China as well as in
Korea and Taiwan. A number of variety of fishes serve as the second- intermediate host of the trematode and these are listed by Komiya (26). The
deeply rooted practice of consuming raw fish in certain ethnic communities
in FE Asia has important bearing on the transmission pattern and epidemiology of clonorchiasis. In Taiwan, the Hakka people, originating from
Kwangtung Province, mainland China, customarily eat raw fresh water fish.
Moreover, the affluent population of this country has a preference for
inadequately cooked fish which is regarded a delicacy. In the mainland
China, the Cantonese people have a more marked preference for the raw fish
in their diet than the people in northern or central China. Raw fishes
are usually served in social gathering of males as an accompaniment to
drinking rice wine in Korea and most Japanese have liking for the slices
of raw flesh of fish treated with soysauce or vinegar.

As the trematodes arrive in the distal bile ducts, these initiate
inflammatory response leading to hyperplasia of bile duct epithelium and
to papillomatous or adenomatous changes. These changes are accompanied by
desquamation of the biliary epithelium. In chronic infections, there is
periepithelial and periportal fibrosis. The extent of hepatic damage is
related to the intensity of infection since the worms may be present in
thousands.C. sinensis infection is also known to be a predisposing factor
for the pathogenesis of cholangio-carcinoma in FE Asian countries (25).

2.2.Opisthorchis viverrini infection (Opisthorchiasis)

The geographical distribution of opisthorchiasis due to O. viverrini
is limited to northeast, north and central Thailand. About seven million
Thai people are affected with the disease to-date and the infection is
a major public health problem (20). Dogs and cats serve as the main
reservoir hosts for the transmission of the infection to man.

The operculate snails, Bithynia siamensis, is the first intermediate
host and the cyprinoid fresh water fishes serve as the second intermediate
host of O. viverrini. The metacercaria are harboured by the fishes which
are the source of infection to mammalian host. In northeastern part of Thailand   a dish, Koi-Pla, consisting of chopped raw flesh of small fishes
and containing condiments, is a very popular food consumed by the rural
population almost daily. This food is an important source of opisthorchiasis in humans. The pathological alterations in the liver due to
O. viverrini infection are described by Harinasuta, Riganti and Bunnag
(19) and these are , in general, identical to those caused by C. sinensis
infection.

2.3. Paragonimus westermani infection (Paragonimiasis)

P.westermani occurs in many countries in SE and FE Asia such as Korea, Japan, Taiwan, Sri Lanka, India, Malaysia, Indonesia, the Philippines, Thailand and northern China and is the well known cause of the important zoonosis, lung-fluke disease, in some of these countries. Large and small forms of P.westermani are reported. Paragonimiasis in Korea, Japan and Taiwan, the countries well known for the human disease, is caused by the large form of the trematode, which reproduce parthenogenetically with triploid chromosomes, while in other Asian countries the small form of the fluke, which reproduce bisexually and has diploid chromosomes, is involved. The large form of the fluke has been given the name, P.pulmonalis, to differentiate it from P.westermani (24).In Thailand, the disease in man is caused by P.westermani and P.heterotremus although few other species of the genus are found in animals (52).Paragonimiasis in the Philippines is caused by P.philippinensis though until about a decade back this trematode was referred to as P.westermani (10).

The definitive hosts of P.westermani other than man include leopard, cat, wild cat, civet cat, tiger, panther, fox, wolf, dog, mongoose and field rats. The molluscs of the families Pleurocercidae and Thiaridae, viz., Semisulcospira libertina (in Japan, Korea, Taiwan and China), Brotia asperata (in the Philippines) and B.costula episcopalis (in Malaysia) are the first intermediate-host of the trematode. The infective metacercariae are found in various species of crabs and cray-fishes (Eriocheir spp., Potamon spp.,Potamiscus spp., Parathelphusa spp., Siamthelphusa spp. etc), which serve as the second intermediate host and sources for infection in definitive hosts. E.japonicus, E.sinensis are the important transmitters of the disease in Korea.

The manner in which the crustaceans are consumed by the local population helps in acquisition of human paragonimiasis. In Korea, raw fresh-water-crabs emersed in soysauce are eaten and cray-fish juice is given to children affected with measles. Wild pigs may serve as a paratenic host in that the meat of this animal harbouring immature P.westermani was found infective for man in Japan. In this country, the infection may be acquired through hands and utensils contaminated with metacercaria during the process of preparation of crab soup. In China, raw crabs sooked in wine or brine are important food and sources of human paragonimiasis. Similarly, in Thailand raw cray-fish salad and raw crab-sauce are important sources of infection. In the Philippines, crab juice is used in certain delicacies and infection is acquired through contamination. In recent years the incidence of human paragonimiasis has drastically reduced in some countries, especially Taiwan, because of the awareness of the masses about the sources of infection.

The flukes are found in lung parenchyma where these provoke leucocytic infiltration followed by development of a broad layer of diffuse connective tissue cyst membrane. The parasite cysts are about one cm in diameter and usually contain two parasites each in thick purulent fluid mixed with blood and golden-brown parasite eggs. The parasite cysts may also be found in ectopic sites like liver, intestinal wall, muscles, brain, peritoneum, pleura etc. or in the subcutaneous tissue abcesses.

110

2.4. Schistosoma japonicum infection (Schistosomiasis japonica)

Schistosomiasis due to S.japonicum is most exhaustively studied and documented since this is an important zoonotic infection in the FE Asian countries. Excellent accounts on the parasite and the disease are available in more recent reviews and books (4,39,54). It is not the intension to deal with this infection in greater detail here and the readers might like to consult these reviews for further reading. However, some of the salient features of schistosomiasis japonica are given below.

China is the largest area of endemecity of S.japonicum; the infection being present in the province around Yangtze river basin and delta and around Taiku, Poyang and Tung-Ting lakes. With the successful control measure programme in China, the number of infected cases has reduced in recent years. In Japan the infection used to be endemic in the island of Honshu (Katayama district, Hirochima and Okayama prefectures; Kofu basin, Yamanashi prefecture; Numazu district, Shizouka prefecture; Tone river basin, Chiba, Saitama and Ibaraki prefectures). In the island of Kyushu, the Chikugo river basin of the Fukuoka and Saga prefectures is endemic for the infection. In the Philippines, endemic areas of the infection are identified in the Irosin Valley on the Luzon island, around lake Naujan on the Mindoro island, in most part of Samar island, east coast of Leyte island and in most part of Mindanao island. In Indonesia, the disease appears confined in central Sulawesi around lake Lindu and the Napu valley.

A large number of mammalian species serve as reservoir hosts for human S.japonicum infection and play an important role in its epidemiology. In China, mouse, dog, goat, rabbit, cow, guinea-pig, sheep, rat, pig, horse and water buffalo harbour the infection; - these are listed in a decreasing order of their transmission potentiality. Various wild mammals including carnivores, rodents, primates, insectivores and artiodactyles are also mentioned to constitute sources of infection in this country under natural conditions. Various rodent species were found to carry this infection in Japan. In the Philippines, dog, cow, pig, rat, carabao and goat are important animal reservoirs. In Sulawesi a variety of rats, wild deer, wild pig, civet cat, cow and dog are infected with S.japonicum. Schistosomiasis japonica is a notorious zoonosis as a large variety of domesticated stock serve its reservoir host. In any integrated control measure directed towards human schistosomiasis japonica an investigation into the role of these hosts on the parasite population dynamics is vital.

Various species of the operculate amphibious mollusc of the genus Oncomelania are the intermediate-hosts of S.japonicum; O.nosophora serving as the vector in Japan and also in certain parts of China, O.hupensis in China and O.quadrasi in the Philippines.

The main lesions of schistosomiasis japonica in man are found in the liver and intestines. The eggs of the parasite sequestrated in the mucosa or submucosa of the intestine initiate granulomatous reaction resulting in formation of pseudotubercles. Similar lesions in liver cause nodular swellings and fibrosis of this organ. Eggs may also be depositted in central nervous system giving rise to intense granulomatous tissue response. Schistosomiasis japonica is mainly an occupational disease of the rural farming communities in FE Asia engaged in paddy fields.

2.5. Schistosoma mekongi infection (Schistosomiasis mekongi)

In more recent years a S. japonicum-like infection was identified in Thailand, Laos, Kampuchea and Malaysia. Since the infection is endemic in the inhabitants of Khong island and among the inhabitants living in the floating villages on the Mekong river south of this island, the parasite received its name as S. mekongi. Mekong schistosomiasis is reviewed by Schneider (49) and its various aspects are dealt with in a multiauthored book edited by Bruce & Sornmani (7). The eggs of S. mekongi are smaller and more subspherical than those of S. japonicum and the prepatent period in the definitive host is longer. The aquatic mollusc, Tricula aperta, is the intermediate-host of S. mekongi and the dogs serve as the reservoir hosts of this infection in the endemic areas. Schistosomiasis japonica and mekongi are the only two zoonotic trematodiasis of SE and FE Asian countries which are non-food borne and the infection is acquired percutaneously through active penetration of the cercariae.

3. INTESTINAL ZOONOTIC TREMATODIASIS
3.1. Trematodiasis with pigs as reservoir hosts:

Few of the trematode infections occurring in the intestine of pigs of SE Asian countries constitute zoonoses and may sporadically threaten human health (28). Three such infections rank high for consideration as following.
3.1.1. Fasciolopsis buski infection (Fasciolopsiasis). Pigs and human are reported to harbour F. buski infection in many of the South Asian countries. Chandler, as quoted by Buckley (8), found that six per cent of the 100 human subjects examined at Manipur Valley, Assam harboured F. buski infection. Buckley's (loc.cit.) own observations further substanciated that 59.7 per cent of the population in Kamrup, Assam carried this infection.

Manning and Ratanarat (36) had shown that the pigs are the reservoir host of F. buski in the central Thailand where an estimated 100,000 persons were infected with this trematode. Thirty per cent of the pigs in Uttar Pradesh, India harboured F. buski (51) while the incidence of human infection in this province was 22.4 per cent (12). A sample of the slaughter pigs examined in Kwangtung, China showed that 10 per cent were infected. In two villages near Dacca,Bängladesh, 39.2 and 8.6 per cent of the children harboured this infection (41) and in southern Taiwan, 19 per cent of the children from seven villages were infected. In one of these villages an infection rate as high as 61 per cent was recorded (31). Newer endemic foci of F. buski infection in human were detected in Thailand (50) and in Maharastra, India (35).

The life-cycle of F. buski is known and the snails Segmentina(Polypylis) hemisphaerula, S. (Trochorbis) trochoideus and Gyraulus chinensis serve as the intermediate-host in Thailand (36), S. calathus, Hippeutis cantori and G. convexiusculus in China while Helicorbis coenosus has been demonstrated to serve as the intermediate-host under experimental conditions in India (51). The cercariae emerging from these snails encyst on aquatic plants like Ipomoea, Trapa, Nymphaea etc. Pigs and human acquire infection by ingesting these aquatic plants containing metacercariae.

The pathogenic effects of F.buski infection in man is not clearly known. Although Plaut et al.(44) have speculated that the trematodes in less than massive number may not be responsible for overt clinical disease, Areekul et al. (3) have found a significant lowering in the serum vitamin B12 content in 100 infected children than that of 60 normal controls and the absorption of this vitamin was impaired in three of the nine patients studied. Further, Chandra (12) has pointed out that anaemia was more frequently encountered among infected individuals and about 85 per cent of the infected cases showed an eosinophil count of more than five per cent. The pathogenicity of fasciolopsiasis, in any case, is influenced by the intensity of infection of the host and a massive worm burden would almost certainly result in adverse effects.

3.1.2. Gastrodiscoides hominis infection. G.hominis infection of pig and man is reported from the northern India, the Philippines, Burma and Thailand and is believed to have a wider geographical range. Buckley (8) reported a very high infection rate in human at Kamrup, Assam where about 41 per cent of the 221 stool samples, mostly from children, were positive for the infection. He, in the year 1964 (9) further emphasised that man is a natural host of G.hominis since the infection at Kamrup was maintained in human population in the absence of a reservoir animal host. There are records to show that G.hominis occurs in monkeys and apes as well (21, 43).

The life-history of G.hominis has been elucidated by Dutt and Srivastava (18) and the tiny aquatic snail Helicorbis coenosus serves as the intermediate-host of the parasite under experimental conditions. Dutt and Srivastava (18) have found that 27 per cent of the 233 slaughter pigs at Bareilly, India were infected with this amphistome and that in 50 per cent of the cases this infection occurred together with F.buski. G.hominis and F.buski use the same molluscan intermediate-host in India and the findings of Buckley (8), showing a high infection rate of these flukes in the same population in Assam, suggest that the two infections may coexist in nature because of their idential transmission pattern.

The mollusc, H.coenosus, abound frequently in the water reservoirs around the pigsties where both these trematode infections are known to be endemic. Intensive search to find H.coenosus naturally infected with G.hominis or F.buski in these waters were, however, not successful (unpublished observations of the author).

Varma (53) had differentiated the parasites found in man and pig as two different strains but this view is generally not accepted.

The pathology and symptomatology of G.hominis infection is uncertain. It is stated that in man the parasite causes inflammation of the mucosa of caecum and ascending colon and diarrhoea. Deaths among untreated patients, especially the children, have been attributed to this infection. Ahluwalia (1) has described a subacute inflammation of the caecum of pigs where the parasite caused desquamation of the mucosa and infiltration with eosinophils, lymphocytes and plasma cells. He has ascribed these lesions to the continual impact of the discoidal region of the parasite on the mucosa of the caecum.

3.1.3. Artyfechinostomum malayanum infection. Leiper (32) described an echinostome trematode from man in the then Federal Malay State under the name Euparyphium malayanum. Four years later, Lane (30) reported another echinostome, Artyfechinostomum sufrartyfex, from a girl in Assam, India. Many of the workers consider these two parasites identical as A.malayanum and Mohandas (37), based on life-history studies and the ensuing evidences, has proved this. A.malayanum also occurs in the small intestine of pigs in these two countries and has been recorded in man in Thailand and Sumatra. Two outbreaks of Paryphostomum sufrartyfex (= A. malayanum) infection among pigs with mortalities among piglets have been reported in Bengal (5). Rarely dogs and rats are infected with this parasite.

The taxonomic position of A.malayanum has aroused considerable controversy for a long time. Mukherjee and Ghosh (40) and Mohandas (38) have rejected the validity of the genus Artyfechinostomum in favour of Echinostoma and have considered A.sufrartyfex a synonym of Echinostoma malayanum. Premvati and Pande (45) have, however, retained the validity of the genus Artyfechinostomum but suppressed, inter alia, its type species, A.sufrartyfex, in favour of A.malayanum. They have considered the latter as the type species of the genus and A.mehrai a synonym of A.malayanum. In the literature A.malayanum is also referred to as Paryphostomum sufrartyfex.

The fresh-water snail, Indoplanorbis exustus serves as the intermediate-host of A.malayanum (17, 33, 37). The cercaria encysts in the snail or the other individual of the same species or the snails of the other species like Lymnaea, Pila, Gyraulus etc. to produce metacercaria. The frog, Rana cyanophlyctis, also serves as the second intermediate-host of the echinostome under experimental conditions in which case the metacercariae are found in the kidney (42). The definitive host acquires infection by ingestion of the snails or frogs harbouring metacercariae.

Generally isolated but grave human cases of A.malayanum infection are recorded in literature as case reports and no clear account on the endemism of the infection is available. Reddy and Varmah (48) reported on a patient, native of Madras, India who died due to malnutrition and aneamia and several thousands of P.sufrartyfex (= A.malayanum) were collected from the small intestine on post-mortem. A very comparable clinical picture of an autopsy finding has been presented by Reddy et al. (47) on a south Indian female who harboured massive A.mehrai (= A.malayanum) infection. Lie and Virik (33) reported E.malayanum (= A.malayanum) infection in a child who was often given Pila snails to eat. Kaul et al. (22) mentioned of a case of intestinal perforation in man due to A.mehrai (= A.malayanum) infection and many worms were recovered in the fluid accumulated in the peritoneal cavity. This patient also ate snails habitually abounding in the paddy fields.

It can be easily conceived that the light infections of A.malayanum showing subclinical parasitism might be more prevalent especially in areas where the incriminated snails are habitually consumed by the population. Such cases of subclinical parasitism without well defined syndromes might easily escape detection by the clinicians.

114

3.2. Trematodiasis with dogs and cats as reservoir hosts :
Few tiny digenetic trematodes of the family Heterophyidae which occur in FE Asian countries in dogs, cats and other fish eating carnivores are transmissible to man. The infection in man is acquired through ingestion of raw or inadequately cooked fish which harbour the infective metacercariae.
3.2.1. Heterophyes heterophyes infection. Several foci of human H.heterophyes infection are known in Japan, central and south China, Korea, Taiwan and the Philippines. In Sino-Japanese area Cerithidea cingulata microptera is the molluscan host and Mugil japonicus is the fish intermediate host. As the raw and pickled fishes are favourable articles of diet in FE Asian countries, infection is acquired through these fishes. Dogs, cats, foxes and probably other fish-eating mammals are the reservoir hosts.

The trematodes are found attached to or burrow into the mucosa of jejunum and ilium and produce mild inflammatory reaction at the sites of their attachment. Heavy infection may result in superficial necrosis of mucosa with significant diarrhoea. The eggs of the trematode may be filtered through the mucosa and be carried to cardiac valves and myocardium. The ova in these ectopic sites produce myocarditis. Up to 15 per cent of fatal myocarditis in the Philippines are related to this infection (23).
3.2.2. Metagonimus yokogawai infection. M.yokogawai infection is frequently found in Korea, Japan and China. Together with clonorchiasis and paragonimiasis, this infection is the most important public health hazard due to trematodiasis in Korea where about 1.2 per cent of the population harbour this infection and in some villages the incidence may reach as high as 20 per cent (13).

The snail hosts of this trematode are Semisulcospira spp. and the infective metacercariae are found in the fresh-water cyprinoid and salmonoid fishes. In Korea, Plecoglossus altivelis is the more common fish intermediate-host. Dogs, cats and rats are naturally infected in endemic areas and constitute sources of infection for man. Pathogenesis of M.yokogawai infection is identical to H.heterophyes infection and the extent of pathological manifestation is invariably related to the intensity of infection.

3.3. Intestinal trematodiasis due to other echinostomes:
3.3.1. Echinostoma ilocanum infection.
Human infections with this trematode are found in Indonesia (Java, Sulawesi) and the Philippines; particularly the Ilocana population of Luzon Province. Naturally infected dogs and rats are reservoir hosts for this infection.

Gyraulus convexiusculus and Hippeutis umbilicatus are the molluscan vector and the cercariae emerging from these snails penetrate and encyst in other fresh water molluscs including Pila spp. and Viviparus spp. These two latter molluscs are favourite delicacies as food items in the endemic areas and man acquire infection by eating these molluscs raw. The infection causes little or no intestinal disturbances though in heavy infections abdominal pain and diarrhoea may ensue due to enteritis.
3.3.2. Echinostoma lindoense infection.
Up to 1956, heavy infections with E.lindoense were found in men living in the surroundings of Lake Lindoe, Sulawesi (Indonesia). The infection was contracted by eating the bivalves, Corbicula lindoensis which harboured the infective metacercariae. However, it has almost disappeared in the

human population of this area possibly because of the change in diet from
bivalves to edible fishes which have been introduced in the Lake Lindoe
(16) though the infection is still prevalent in the animal reservoirs (11).
Besides the bivalves, Viviparus javanicus also harboured the metacercariae.

4. PRAZIQUANTEL; A PANACEA FOR ZOONOTIC TREMATODIASIS

Modern drugs have largely superseded the older ones because of their
broader spectrum of activity and reduced toxicity and, in this context, the
development of praziquantel is a brilliant example of this phenomenon and a
major step forward for the chemotherapy of zoonotic trematodiasis. In a recent review (6) the efficacy of this drug against various trematode infections of man is illustrated. For the treatment of zoonotic trematodiasis
of SE and FE Asian countries the following drug regimens are considered
feasible (Table 1).

TABLE 1.	TREMATODIASIS	vs	PRAZIQUANTEL
	Regimen		Follow up cure rates (%)
Schistosomiasis (japonica/mekongi)	Single day, 2 x 30 mg/kg		87.5
Clonorchiasis/ opisthorchiasis	Single day, 3 x 25 mg/kg		81 - 97
Paragonimiasis	2 consecutive days, 3 x 25 mg/kg		85.7
Fasciolopsiasis	Single dose, 15 mg/kg		100
Metagonimiasis	Single day, 2 x 20 mg/kg		100

It appears certain that the spectrum of activity of praziquantel against
rare trematode infections is not fully understood. Four hundred and eleven
Thai patients were treated with praziquantel for supposedly O.viverrini
infection and some of these patients were purged following chemotherapy.
Besides O.viverrini, the rare trematodes recovered in the stool of these
patients included Prosthodendrium molenkampi, Phaneropsolus bonnei,
Haplorchis spp. etc. The eggs of these rare trematodes are very much identical to O.viverrini and complicate specific diagnosis of opisthorchiasis
(46). Although the question of chemotherapy of zoonotic trematodiasis may
have been resolved to a great extent, other comprehensive and integrated
control measures are indespensable in achieving a substancial reduction
in the incidence of human trematodiasis.

ACKNOWLEDGEMENT
The author is grateful to Mrs.Paula Dumoulin for the assistance.

REFERENCES

1. Ahluwalia SS: Gastrodiscoides hominis (Lewis and McConnell) Leiper, 1913; the amphistome parasite of man and pig. Indian Journal of Medical Research, 48: 315-325, 1960.
2. Arambulo III PV and Moran N: Food transmitted parasitic zoonoses; sociocultural & technological determinants. International Journal of Zoonoses, 7: 135-141, 1980.
3. Areekul S, Jaroonvesama N, Charoenlarp K, Kasemsuth R and Cheeramakara C: Serum vitamin B_{12} and folic acid levels in patients with fasciolopsiasis.Southeast Asian Journal of Tropical Medicine and Public Health, 10: 67-72, 1979.
4. Beaver PC, Jung RC and Cupp EW: Clinical Parasitology. Lea & Febiger, Philadelphia, 1984.
5. Bhattacharyya HM, Das SK, Sinha PK, Biswas SN and Bose PK: Mortality in pig due to Paryphostomum sufrartyfex (Lane, 1915) 49: 976-978, 1972.
6. Brandt JRA, Kumar V and Geerts S: Praziquantel In: The Antimicrobial Agents (P.K., Peterson and J.Verhoef, Eds), Elsevier Science Publishers B.V.,Amsterdam: 287-294, 1986.
7. Bruce JI and Sornmani S: The Mekong Schistosome. Whitmore Lake, Michigan, Malacological Review Supplement, 1980.
8. Buckley JJC: Observations on Gastrodiscoides hominis and Fasciolopsis buski in Assam. Journal of Helminthology, 17: 1-12, 1939.
9. Buckley JJC: The problem of Gastrodiscoides hominis (Lewis and McConnell, 1876) Leiper, 1913. Journal of Helminthology, 38: 1-6, 1964.
10. Cabrera BD: Paragonimiasis in the Philippines: Current status. Drug Research, 34: 1188-1192, 1984.
11. Carney WP, Sudomo M and Purnomo: Echinostomiasis: A disease that disappeared. Tropical and Geographical Medicine, 32: 101-105, 1980.
12. Chandra SS: A field study on the clinical aspects of Fasciolopsis buski infections in Uttar Pradesh. Medical Journal of Armed Forces, India, 32: 181-189, 1976.
13. Cho SY, Kang SY and Lee JB: Metagonimiasis in Korea. Drug Research, 34: 1211-1213, 1984.
14. Choi, Dong-Wik: Clonorchis sinensis: Life cycle, intermediate hosts, transmission to man and geographical distribution in Korea. Drug Research, 34: 1145-1151, 1984.
15. Clarke MD, Khaw OK and Cross JH: Clonorchiasis in Sun Moon lake area. Chinese Journal of Microbiology, 4: 19-26, 1971.
16. Clarke MD, Carney WP, Cross JH, Hadidjaja P, Oemijati S and Joesoef A: Schistosomiasis and other human parasitoses of lake Lindu in Central Sulawesi (Celebes), Indonesia. American Journal of Tropical Medicine and Hygiene, 23: 385-392, 1974.
17. Dissamarn R, Aranyakananda P, Srivornath P, Chai-Anan P, Thirapat K and Chitrakorn P: The life-history of Paryphostomum sufrartyfex (Lane, 1915) Bhalerao 1931. Journal of Thai Veterinary Medical Association, 17: 11-16, 1966.
18. Dutt SC and Srivastava HD: The life-history of Gastrodiscoides hominis (Lewis and McConnell, 1876) Leiper, 1913, -the amphistome parasite of man and pig. Journal of Helminthology, 46: 35-46, 1972.
19. Harinasuta T, Riganti M and Bunnag D: Opisthorchis viverrini infection: Pathogenesis and clinical features. Drug Research, 34: 1167-1169, 1984.

20. Harinasuta C and Harinasuta C:Opisthorchis viverrini: Life cycle, intermediate hosts, transmission to man and geographical distribution in Thailand. Drug Research, 34: 1164-1167, 1984.
21. Herman LH: Gastrodiscoides hominis infestation in two monkeys. Veterinary Medicine, 62: 355-356, 1967.
22. Kaul BK Singhal GD and Pillai PN: Artyfechinostomum mehrai infestation with bowel perforation. Journal of Indian Medical Association, 63: 263-265, 1974.
23. Kean BH and Breslau RC: Cardiac heterophyidiasis. In: Parasites of the Human Heart, Grune & Stratton, New York: 95-103, 1964.
24. Kim Dong-Chan: Paragonimus westermani: Life cycle, intermediate hosts, transmission to man and geographical distribution in Korea. Drug Research, 34: 1180-1183, 1984.
25. Kim Yong-Il: Liver carcinoma and liver fluke infection. Drug Research, 34: 1121-1126, 1984.
26. Komiya Y: Clonorchis and Clonorchiasis. Advances in Parasitology, 4: 53-106, 1966.
27. Komiya Y and Suzuki N: In:Progress of Medical Parasitology in Japan (K.Morishita, Y.Komiya and H.Matsubayashi, eds), Vol.1: 551-645. Meguro Parasitological Museum, Tokyo, 1964.
28. Kumar V: The digenetic trematodes, Fasciolopsis buski, Gastrodiscoides hominis and Artyfechinostomum malayanum as zoonotic infections in southeast Asia. Annals de la Société Belge de Médecine Tropicale, 60: 331-339, 1980.
29. Kumar V: Repercussions of traditional food habits on the transmission dynamics of food-transmitted helminthiasis of man. Proceedings of the International Colloquium on Tropical Animal Production for the Benefit of Man, Antwerp: 539, 1982.
30. Lane C: Artyfechinostomum sufrartyfex. A new parasitic echinostome of man. Indian Journal of Medical Research, 2: 977-983, 1915.
31. Lee HH: Fasciolopsis buski infection among children in Liu-Ying Primary School in Taiwan Hsien, South Taiwan. Chinese Journal of Microbiology, 5: 110-114, 1972.
32. Leiper RT: A new echinostome parasite in man. Journal of London School of Tropical Medicine, 1: 27-28, 1911.
33. Lie KJ: The life-history of Echinostomum malayanum Leiper, 1911. Tropical and Geographical Medicine, 15: 17-24, 1963.
34. Lie KJ and Virik HK: Human infection with Echinostomum malayanum Leiper, 1911 (Trematoda: Echinostomatidae). Journal of Tropical Medicine and Hygiene, 66: 77-82, 1963.
35. Manjarumkar PV and Shah PM: Epidemiological study of Fasciolopsis buski in Palghar Taluk. Indian Journal of Public Health, 16: 3-6, 1972.
36. Manning GS and Ratanarat C: Fasciolopsis buski (Lankester, 1857) in Thailand. American Journal of Tropical Medicine and Hygiene, 19: 613-619, 1970.
37. Mohandas A: Artyfechinostomum sufrartyfex Lane 1915, a synonym of Echinostoma malayanum Leiper, 1911 (Trematoda: Echinostomatidae). Acta Parasitologica Polonica, 19: 361-368, 1971.
38. Mohandas A: The present status of the genera Artyfechinostomum, Neoartyfechinostomum and Pseudoechinostomum and the validity of the species included in the three genera (Trematoda: Echinostomatidae). Rivista di Parassitologia, 35: 205-212, 1974.

39. Mott KE: S.japonicum and S.japonicum-like infections. In: Schistoso-miasis; Epidemiology, Treatment and Control (P.Jordan and G.Webbe, eds). William Heinemann Medical Books Ltd, London: 128-149, 1982.

40. Mukherjee RP and Ghosh RK: On the synonymy of the genus Artyfechinos-tomum Lane, 1915 (Trematoda: Echinostomatidae). Proceedings of the Indian Academy of Sciences, 68: 52-58, 1968.

41. Muttalib MA and Islam N: Fasciolopsis buski in Bangladesh; a pilot study. Journal of Tropical Medicine and Hygiene, 78, 135-137, 1975.

42. Nath D: Rana cyanophlyctis as the second intermediary of Artyfechinos-tomum sufrartyfex (Echinostomatidae: Trematoda). Current Science, 38 : 342-343, 1969.

43. Pester FRN and Keymer IF: Gastrodiscoides hominis from an orangutan, Pongo pygmaeus. Transactions of the Royal Society of Tropical Medicine and Hygiene, 62: 10-11, 1968.

44. Plaut AG, Kampanartsanyakorn C and Manning GS: A clinical study of Fasciolopsis buski infection in Thailand. Transactions of the Royal Society of Tropical Medicine and Hygiene, 63: 470-478, 1969.

45. Premvati G and Pande V: On Artyfechinostomum malayanum (Leiper, 1911) Mendheim, 1943 (Trematoda: Echinostomatidae) with synonymy of allied species and genera. Proceedings of the Helminthological Society of Washington, 41: 151-160, 1974.

46. Radomyos P, Bunnag D and Harinasuta T: Worms recovered in stools fol-lowing praziquantel treatment. Drug Research 34: 1215-1217, 1984.

47. Reddy DB, Ranganaykamma I and Venkataratnam D: Artyfechinostomum mehrai infestation in man. Journal of Tropical Medicine and Hygiene, 67: 58-59, 1964.

48. Reddy RG and Varmah K: Paryphostomum sufrartyfex (intestinal fluke) infection in man. Indian Medical Gazette, 85: 546-547, 1960.

49. Schneider CR: Schistosomiasis in Cambodia: A review. Southeast Asian Journal of Tropical Medicine and Public Health, 7: 155-165, 1976.

50. Soavakontha S, Charoenlarp P, Radomyos P and Harinasuta C: A new ende-mic area of Fasciolopsis in Thailand. The Proceedings of the 1st International Congress of Parasitology, Rome: 21-26, 1966.

51. Tripathi JC, Srivastava HD and Dutt SC: A note on experimental infec-tion of Helicorbis coenosus and pig with Fasciolopsis buski. Indian Journal of Animal Science, 43: 647-649, 1973.

52. Vanijanonta S, Bunnag D and Harinasuta T: Paragonimus heterotremus and other Paragonimus spp.in Thailand: Pathogenesis, clinic and treatment. Drug Research, 34: 1186-1188, 1984.

53. Varma AK: Human and swine Gastrodiscoides. Indian Journal of Medical Research, 42: 475-479, 1954.

54. Webbe G: The intermediate hosts and host parasite relationships. In: Schistosomiasis; Epidemiology, Treatment and Control (P.Jordan and G.Webbe, eds). William Heinemann Medical Books Ltd, London: 16-49, 1982.

Observations on human and animal schistosomiasis in Senegal

D. Rollinson, J. Vercruysse*, V.R. Southgate, P.J. Moore °, G.C. Ross, T.K. Walker and R.J. Knowles

Abstract

The main features that distinguish Schistosoma curassoni from S.haematobium and S. bovis are considered. Studies relating to egg morphology, surface structure, growth studies, egg-production, intermediate hosts, enzymes, ribosomal RNA and definitive host specificity clearly show the validity of S. curassoni. Results are reported of a schistosomiasis survey, which was designed to investigate the role, if any, of S.curassoni as a zoonosis, in the area of Tambacounda and Sintiou Maleme, Senegal. At Tambacounda abattoir, schistosome worms were recovered from 19 cattle out of 46 examined; 11 sheep and 12 goats were examined but only one goat was found with a light infection. Subsequent enzyme and morphological analyses showed 18 of the cattle to be infected with S. curassoni. S. bovis was observed in one animal and suspect hybrid worms were recovered from two animals. B. umbilicatus was collected from water bodies at Sintiou Maleme and Pigna. Of 310 snails from Pigna, 22 shed schistosome cercariae whereas, 4 out of 344 snails were positive from Sintiou Maleme. Both S. curassoni and S. haematobium were isolated from wild-caught snails. Examination of 58 urines and 40 stool samples from children in Sintiou Maleme revealed 42 cases of urinary schistosomiasis; only one child was found to be passing S. haematobium-like eggs in faeces. Ten isolates were established in the laboratory, 9 from eggs in urine and 1 from eggs in stools, all were identified as S. haematobium. No evidence was found to support the view that S. curassoni develops in man.

1.Introduction

In Senegal, three species of Schistosoma are recognised as belonging to the S. haematobium group due to their ability to develop in species of the freshwater snail Bulinus and by the possession of terminal-spined eggs. S. haematobium is an important human pathogen, common throughout much of Africa and the Middle East, which is associated primarily with the vesicular veins of the bladder. S. bovis

Geerts, S., Kumar, V., Brandt, J. (eds) Helminth Zoonoses
© 1987 Martinus Nijhoff Publishers, Dordrecht. ISBN 0-89838-896-1. Printed in the Netherlands.

120

is of veterinary importance being a common parasite of ruminants in parts of North, West, East and Central Africa, the Middle East and the Mediterranean region and frequents the mesenteric veins of the intestines. Similarly, S. curassoni, which has recently been redescribed (18), is a parasite of ruminants found mainly in the mesenteric veins but appears to be confined in its distribution to parts of West Africa. As yet, it is not clear how these three species interact in nature and clarification is needed as to whether S. curassoni infects man.

In contrast to the often close relationship of schistosomes with their intermediate snail hosts, the definitive host range of some schistosome species extends to a variety of mammalian species. Knowledge of the definitive host range of a species is of the utmost importance from an epidemiological viewpoint. In order to identify clearly the species or strain of Schistosoma responsible for infection it is often necessary to isolate the parasite and examine various features in detail, especially if distinctive morphological characters of the schistosome egg are lacking. Our aim in this paper is to consider in detail the features of S. curassoni which facilitate its recognition as a discrete entity from S. haematobium and S. bovis in Senegal. In a further attempt to elucidate the host-specificities of S. curassoni, data pertaining to isolates from man, cattle and snails collected during a small survey in the area of Sintiou Maleme and Tambacounda,(fig 1) Senegal, are also presented.

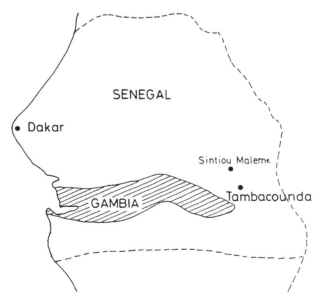

Figure 1. Sketch map of Senegal detailing the position of Tambacounda and Sintiou Maleme

2.Historical Background

Belief in the existence of S. curassoni has waxed and waned over the last 55 years. The original collection of adult schistosomes was made by M.G. Curasson from the mesenteric veins of a domestic cow originating from around the area of Bamako, Mali. Utilising 10 males and 2 females, Brumpt described the schistosome as S. curassoni (3). The name was soon challenged; in 1932, Bhalareo (2) considered S. curassoni to be synonymous with S. bovis, while in 1961, LeRoux (11) argued for synonymy with S. mattheei or S. bovis. A year later, Grétillat (7) reported the presence of S. curassoni in the mesenteric veins of cattle and sheep in Senegal and Mauritania. Furthermore, he created the genus Proschistosoma to take account of a peculiar mode of asexual reproduction, multiplication by asexual budding of sporocysts, that he had observed in the snail host (7). Most importantly, he claimed to infect sheep with schistosome cercariae from snails that had been infected with miracidia originating from urine samples of children from Dakar and Kiréne (9). The schistosomes were identified as P. curassoni and, for the first time, the question as to whether S. curassoni was a zoonosis was raised (9,10). Shortly after these findings were reported, the validity of S. curassoni was again challenged. Capron and his colleagues (4) made a comparative study of the development of four isolates of S. haematobium from Algeria, Mauritania, Morocco and Senegal together with material provided by Grétillat. The conclusion based on their data was that the genus Proschistosoma should be declared invalid and that S. curassoni should be considered as a synonym of S. bovis. Independently and at the same time, Pitchford (13) put forward the view that S. curassoni should be declared synonymous with S. mattheei until more material was available for study. For the next 19 years, authors reviewing the literature on schistosomes tended to disregard S. curassoni (6,12) or considered it to be invalid (5). However, Vercruysse, Southgate and Rollinson (18) redescribed S. curassoni from sheep and goats in Senegal and demonstrated certain criteria for distinguishing this parasite from S. haematobium, S. bovis and S. mattheei ; they also presented results of infection experiments that suggested it was unlikely that S. curassoni was a zoonosis. In contrast, Albaret and his colleagues (1) formed rather different conclusions using two morphological indices; one was based on the relative position of sensory receptors of cercariae and the other on the ratio of length/width of eggs. They confirmed the presence of a parasite, S. curassoni in Senegal, which differed from S. haematobium and S. bovis, but their preliminary evidence led them to conclude that S. curassoni occurs in man.

3.Distinguishing features of S. curassoni

3.1 Egg morphology : Given the past confusion concerning the identity of S. curassoni it is perhaps surprising that egg morphology can be used to separate clearly this species from both S. bovis and S. mattheei (17) . The eggs of S. curassoni measure 149.4 ± 13.2 µm long and 62.8 wide ± 4.9 µm; in contrast to the eggs of S. bovis which measure 223.9 ± 13.2 µm long, 66.0 ± 5.5 µm wide and to those of S. mattheei which measure 173 µm long, 53 µm wide (12). The eggs are, however, similar to those of S.haematobium which measure 153.1 ± 11.1 µm long and 62.4 ± 12.1 µm wide.

3.2 Surface structure : Scanning electron microscopical studies of adult male worms have shown that S. bovis can be easily distinguished from both S. curassoni and S. haematobium on the basis of the structure of the tubercles (16). The tubercles on the dorsal and dorso-lateral surface of male S. bovis are devoid of spines, whereas spines are present on the tubercles of male S. curassoni and S. haematobium.

3.3 Growth studies and egg production : Growth studies of adult worms clearly show that S. curassoni is a much larger worm than S. haematobium. In hamsters, the cross-over point for S. curassoni (i.e. the point at which the average lengths of the paired adult male and female worms are the same) was 13.7 mm at 42 days post-infection, whereas it was 8.3 mm at about 62 days post-infection for S. haematobium. The development times and sizes of S. bovis are roughly comparable with S. curassoni. The mean average production of 167 eggs/day per female worm of S. curassoni in hamsters was much greater than either that of S. bovis at 95 eggs/day or S. haematobium at 86 eggs/day. Eggs were observed in the uteri of 60% of the paired S. curassoni at 40 days post-infection, similar to S. bovis, whereas at 60 days post-infection only 22.2% of the S.haematobium worms were gravid (15).

3.4 Intermediate hosts : Snail infection experiments have shown S. curassoni to be compatible with Bulinus umbilicatus, marginally so with B. senegalensis, and incompatible with B. guernei. Isolates of S. haematobium from Senegal exhibit similar preferences being compatible with B. umbilicatus and B. senegalensis but not B. guernei. Interestingly, S. bovis appears to be compatible with all these species (15).

3.5 Enzymes : Analyses of adult worms by isoelectric focusing have revealed enzyme systems which are useful for differentiating the three species. Representative isolates of each of the species from Senegal have shown intra - population heterogeneity which was probably a reflection of the fact that for S. bovis and S. curassoni adult worms from natural infections of ruminants collected at slaughter were analysed rather than laboratory passaged material.

S.bovis can be differentiated from S. haematobium by patterns of glucose-6-phosphate dehydrogenase (G6PDH), hexokinase (HK), acid phosphatase (AcP) and small differences in glucose phosphate isomerase (GPI). Similarly S. bovis and S. curassoni can be distinguished by the different patterns of phosphoglucomutase (PGM), GPI, HK and AcP. Greater similarity appears to exist between S. haematobium and S. curassoni but the two species can be clearly differentiated by patterns of PGM and HK (15).

3.6 DNA : Recently, the use of cloned ribosomal RNA gene probes to differentiate S. haematobium from related species has been reported (20). The schistosome rRNA gene unit consists of a regular interspersion of the two genes encoding the large and small rRNA units with two spacers. The large spacer region is not transcribed while the small spacer is part of the transcription unit. Differences in the length of the non-transcribed spacer allow the recognition of S. haematobium, the latter has a 0.5 kb deletion in this region which does not occur in either S. curassoni or S. bovis as shown in Figure 2. No clear differences in the rRNA gene unit have emerged between S. bovis and S. curassoni but S. mattheei can be clearly distinguished.

3.7 Definitive host specificity : S. curassoni develops experimentally in sheep, goats and cattle, as does S. bovis and both parasites occur in ruminants in Senegal. In contrast, isolates of S. haematobium did not develop a patent infection in experimental sheep (18). S. haematobium will develop in baboons and a natural infection of a baboon from Senegal has been reported (17). A baboon exposed experimentally to initially 1000 cercariae of S. curassoni and then 7000 cercariae 78 days later did not develop an infection (18).

Thus, there are a number of features by which S. curassoni can be recognised. The identification of parasites within ruminants in Senegal should be facilitated, in the first instance, by examination of the size and shape of the egg and the surface structure of the adult male worms. However, if S. curassoni is capable of infecting man, then the similarity of egg morphology with that of S. haematobium would preclude the possibility of rapid identification of human infections and other features of the parasite as revealed by growth studies, enzyme and DNA analyses would have to be examined to distinguish the two species.

The following section outlines a preliminary survey conducted in an area of natural transmission of S. curassoni, S. haematobium and possibly S. bovis.Isolates from cattle and human populations and naturally infected B.umbilicatus have been identified by the use of enzyme and rRNA analyses in conjunction with morphological features.

4. Schistosomiasis survey in an area of Senegal endemic for S. curassoni and S. haematobium .

The determination of the intermediate host range of S. curassoni in the laboratory, implicating the role of B. umbilicatus , facilitated the identification of natural transmission sites. Our previous isolates of S. curassoni had been obtained from natural infections in animals at slaughter in Dakar (18) and questions persisted as to the origin of the infections in these animals. As part of a survey of human and animal schistosomiasis in the Senegal River Basin, a field study of urinary schistosomiasis in villages along the Basin as far east as Matam documented the presence of S. haematobium in the human population and indicated the importance of B. senegalensis in transmission (19). Apart from an isolated find of S. curassoni in a sheep in the region of Matam (J.V. unpublished), our knowledge of the distribution of S. curassoni was limited.

A further small survey was conducted during November and December, 1985, primarily in the Tambacounda region (Eastern Senegal) (Fig1). The three aims of the survey were as follows :

1) To examine animals at slaughter in local abattoirs in order to pinpoint areas possibly associated with transmission of S. curassoni.

2) To find water bodies utilised by people and their animals which harboured potential intermediate hosts and to identify the parasites involved.

3) To isolate and identify parasites from people who would be expected to be in contact with the suspect transmission sites.

4.1 Parasites from cattle.

At Tambacounda abattoir, examination of the mesenteric veins of 46 cattle revealed that 19 were positive for schistosomes. Eleven sheep and 11 goats were negative for infection, one goat had a very light infection. The worms were collected, rinsed in saline and stored in liquid nitrogen pending subsequent analyses. A breakdown of the identifications based on the analysis of AcP by isoelectric focusing together with the examination of intra-uterine eggs is given in Table 1. AcP patterns of S.curassoni and S.bovis and a possible hybrid are shown in Figure 2. S. curassoni occurred in all the positive animals. In addition, S. bovis was seen in one animal and a male S. bovis / female S. curassoni pair was observed. Suspect hybrid worms were found in two of the animals.

Table 1. Identification of schistosomes recovered from cattle at Tambacounda
abattoir, based on AcP-type and morphological analyses.

Reference No.	S. curassoni M	S. curassoni F	S. bovis M	S. bovis F	Hybrid ? M	Hybrid ? F
40	7	5			1	1
41	4	4				
42	1	1				
43		1	3			
44	1	1				
45		1				
47	1					
48	1	1				
49		2			2	
50	1	1				
51	76	66				
58	1					
59	4	1				
60	1					
61	2	1				
62	7	1				
63	27	22				
64	36	24				

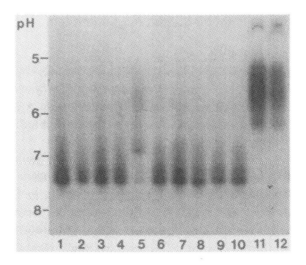

Figure 2. Acid phosphatase patterns of adult male worms of S. curassoni (Lanes 1-4, 6-10),
S. bovis (Lanes 11, 12) and a possible hybrid (Lane 5)

4.2 Parasites isolated from B. umbilicatus

Approximately 25 km to the west of Tambacounda along the road to Koussanar lies the village of Sintiou Maleme. This is an area with large livestock holdings. On the eastern edge of the village, pools were found along a drying-up river bed in which B.umbilicatus were plentiful,together with a few B. forskalii. This site was visited frequently by cattle and also during certain times of the year by the local children. B.umbilicatus was also abundant at a larger alluvial pool some 15-20 km to the north at Pigna. At the time of collection at this site, both animals and people were in contact with the water. The wild caught snails were maintained in the laboratory and monitored for shedding over a 10 week period. Of the snails collected from Pigna, 310 survived and 22 produced schistosome cercariae; of 344 from Sintiou Maleme, 4 were found to be positive. Cercariae from some of these snails were used to infect mice, hamsters and sheep. A sheep exposed to 2000 cercariae from the snails from Sintiou Maleme excreted eggs identifiable as S. curassoni, this was confirmed by AcP analysis of the adult worms. Egg measurements for this isolate recovered from livers of mice were 150.9 ± 10.4 μm long, 65.5 ± 2.6 μm wide. Similarly, a sheep exposed to 8000 cercariae from the 15 snails from Pigna was shown later to harbour S. curassoni worms. However, at least one snail was naturally infected with S. haematobium, as single adult male worms recovered from mice produced patterns of PGM, G6PDH and HK diagnostic of S. haematobium. Additionally, laboratory isolates established with the cercariae from snails collected at Pigna, were found to possess rRNA profiles indicative of S. haematobium and S. curassoni.

4.3   Parasites isolated from man

Urine and stool samples were collected during the mid-morning from local children, in the age group 4-14 years, attending the school at Sintiou Maleme. Of 58 urines examined by egg sedimentation, 42 were found to be positive; 40 stools examined by the Kato technique revealed 1 positive infection with S. haematobium-like eggs.

Miracidia were hatched from some of the urines with the highest egg counts and used to infect batches of B. wrighti , a laboratory snail host susceptible to species within the S. haematobium group. In this way, 9 laboratory isolates were established. Similarly, eggs from two successive stool samples, from the one boy passing eggs in the faeces, were hatched and this parasite was established in culture. A few morphologically similar eggs were also observed in the urine of this child but insufficient viable eggs were found to enable isolation of the parasite.

Cercariae resulting from the snail infections, representing the 10 individual isolates, were used to infect hamsters and mice and the resulting adult worms were analysed paying particular attention to growth and maturation rates, enzyme profiles and the size of the non-transcribed spacer of the rRNA gene unit. The parameters tested showed that all ten of the isolates were S. haematobium; the PGM patterns associated with S. haematobium and six of the isolates, including the one isolated from faeces, were shown to possess the deletion in the non-transcribed spacer typical of S. haematobium.(Figure 3).

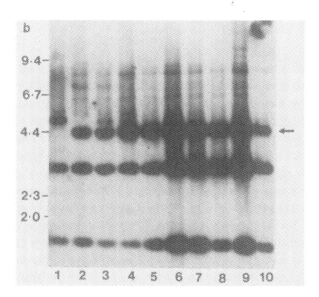

Figure 3. Autoradiograph of the hybridisation of the 32P-labelled PSM889 rRNA probe with ECORI digested genomic DNA from adult worms. Lane 1, S.curassoni; Lanes 2,4,5,7,8,10, S.haematobium isolates established from eggs in human urine; Lanes 6,9, S.haematobium isolated from eggs in human stools; Lane 3, S. haematobium isolated from naturally infected B. senegalensis (Senegal). Band sizes are marked. The largest main band (arrowed) encompasses the non-transcribed spacer of the rRNA gene unit, the smaller size of this fraction is characteristic of S. haematobium.

Particular attention was given to the isolate obtained from eggs in faeces. In ruminants, S. curassoni is associated primarily with the mesenteric veins of the intestines, although some bladder involvement has been reported (20). It, therefore, seemed possible that if the parasite developed in man, it might well be expected to have a preference for a similar site, a possibility previous

128

workers appear to have neglected. However, no differences were found between
this parasite and the 9 isolates from eggs in urine; growth rates, PGM, G6PDH,
AcP and HKprofiles and rRNA analyses were exactly the same as for S.
haematobium. An additional check was made by exposing a sheep to this isolate.
Approximately 1800 cercariae in total on two successive days were used to infect
a sheep but the infection did not become patent and no worms were recovered at
autopsy.

In addition to the children, urines and stool samples were collected from two
men, aged 34 and 40, who worked with cattle. All the samples were negative for
schistosome eggs.

5 .Conclusion

The survey revealed a prevalence of S. curassoni of 41.3% in cattle at
slaughter at Tambacounda abattoir. In the vicinity of Sintiou Maleme and Pigna
two transmission sites were located. Natural infections of both S.curassoni and S.
haematobium in B. umbilicatus were identified. A high prevalence of
schistosomiasis, 72.4% was found in children attending the school at Sintiou
Maleme. Rigorous analyses of 10 isolates (9 from eggs in urine, 1 from eggs in
stools) identified these parasites as S. haematobium.The finding of S.
haematobium eggs in the stools of one child was interesting. This phenomenon is
usually associated with heavy urinary infections but few eggs were found in urine
samples from this child. No evidence has been found, therefore, to support the
idea that S. curassoni develops in man, contrary to the views of others (1).
Negative results of this sort are by their nature somewhat inconclusive but the data
strongly suggest that it is unlikely that S. curassoni is a zoonosis in this area of
Senegal but conclusive evidence must await the analyses of more isolates.

The definitive host specificities of S. haematobium and S. curassoni appear to
be effective barriers to prevent the inter-mixing of these two species. Should the
mechanisms which preclude either parasite from infecting the hosts of the other
break down, then it is likely that quite a different picture of schistosomiasis would
evolve. Some schistosome species of the S. haematobium group will hybridise in
nature, for example, hybridisation of S. haematobium with S. intercalatum and
also S.mattheei is well documented. When species do hybridise, the disease pattern
can change radically over a relatively short period of time (14). S. bovis and S.
curassoni do share the same definitive hosts and our preliminary observations on
natural inter-specific pairings and enzyme profiles suggest that some
hybridisation is occurring between these two species; this may account, in part,
for some of the earlier confusion concerning the status of S. curassoni.

Acknowledgements

This investigation was supported in part by the Parasitic Diseases Programme of the World Health Organisation. We are also grateful to the UNDP/World Bank/WHO Special Programme for Research and Training in Tropical Diseases and the Medical Research Council for financial assistance. We are indebted to Dr B.M. Greenwood, Director , and Drs H.A. Wilkins and P. Hagan of the Medical Research Laboratories in the Gambia for their help and the use of facilities at the field station in Basse. We thank Dr.A.Gueye, Head of the Department of Research, Planning and Formation for allowing us to carry out our investigation in Senegal.

References.

1. Albaret JL, Picot H, Diaw OT, Bayssade-Dufour Ch, Vassiliades G, Adamson M, Luffau G & Chabaud AG : Enquête sur les schistosomes de l'homme et du bétail au Sénégal, à l'aide des identifications spécifiques fournies par la chétotaxie des cercaires. I. Nouveaux arguments pour la validation de S. curassoni Brumpt, 1931, parasite de l'homme et des Bovidés domestiques. Annales de Parasitologie Humaine et Comparée 60: 417-434, 1985.
2. Bhalérao G : On the identity of the Schistosome found in cases of bovine nasal granuloma and some observations on a few other members of the Schistosomatidae. Indian Journal of Veterinary Science and Animal Husbandry, II : 338-356, 1932.
3. Brumpt E : Description de deux bilharzies de mammifères africains, Schistosoma curassoni, sp. inquir. et Schistosoma rodhaini n.sp. Annales de Parasitologie Humaine et Comparée, 9: 325-338, 1931.
4. Capron A , Deblock S, Biguet J, Clay A, Adenis L & Vernes A : Contribution à l'étude expérimentale de la bilharziose à Schistosoma haematobium. Bulletin of the World Health Organisation, 32 : 755-778, 1965.
5. Christensen N O,Mutani A & Frandsen F: A review of the biology and transmission ecology of African bovine species of the genus Schistosoma. Zeitschrift für Parasitenkunde, 69 : 551-570, 1983.
6. Davis GM: Snail hosts of Asian Schistosoma infecting Man : evolution and coevolution. Malacological Review, Supplement 2: 195-238, 1980.
7. Grétillat S: Etude du cycle évolutif du schistosome des ruminants domestiques de l'Ouest Africain et confirmation de l'espèce Schistosoma curassoni Brumpt, 1931. Annales de Parasitologie Humaine et Comparée, 37:556-568, 1962.

8. Grétillat S: Recherches sur le cycle évolutif du schistosome des ruminants domestiques de l'Ouest Africain, (Schistosoma curassoni Brumpt 1931). Comptes Rendus de l'Académie des Sciences, Paris, 255: 1657-1659, 1962.

9. Grétillat S: Une nouvelle zoonose, la 'Bilharziose Ouest Africaine' à Schistosoma curassoni Brumpt, 1931, commune à l'homme et aux ruminants domestiques. Comptes Rendus de l'Académie des Sciences, Paris, 255: 1805-1807, 1962.

10. Grétillat S: Contribution à l'étude de l'épidémiologie des bilharzioses humaines et animales en Haute-Casamance (Sénégal) et en Mauritanie. Revue d'Elevage et Médecine Vétérinaire des Pays Tropicaux, 16: 323-334, 1963.

11. Le Roux PL: Some problems in bilharziasis in Africa and the adjoining countries. Journal of Helminthology, R.T. Supplement : 117-126, 1961.

12. Loker ES: A comparative study of the life-histories of mammalian schistosomes. Parasitology, 87: 343-369, 1983.

13. Pitchford RJ: Differences in the egg morphology and certain biological characteristics of some African and Middle Eastern schistosomes, genus Schistosoma, with terminal-spined eggs. Bulletin of the World Health Organisation, 32: 105-120, 1965.

14. Rollinson D & Southgate VR: Schistosome and snail populations : genetic variability and parasite transmission. In: Ecology and Genetics of Host-Parasite Interactions, D Rollinson & RM Anderson (eds), Academic Press, Linnean Society Symposium Series, 11: 91-109, 1985.

15. Southgate VR, Rollinson D, Ross GC, Knowles RJ & Vercruysse J : On Schistosoma curassoni, S. haematobium and S. bovis from Senegal: development in Mesocricetus auratus, compatibility with species of Bulinus and their enzymes. Journal of Natural History, 19: 1249-1267, 1985.

16. Southgate VR, Rollinson D & Vercruysse J: Scanning electron microscopy of the tegument of adult Schistosoma curassoni and comparison with male S.bovis and S. haematobium from Senegal. Parasitology, 93: 433-442, 1986.

17. Taylor MG, Nelson GS & Andrews BJ: A case of natural infection of S. haematobium in a Senegalese baboon (Papio sp.). Transactions of the Royal Society of Tropical Medicine and Hygiene, Laboratory meeting, 66: 16-17, 1972.

18. Vercruysse J, Southgate VR & Rollinson D: Schistosoma curassoni Brumpt, 1931, in sheep and goats in Senegal. Journal of Natural History, 18: 969-976, 1984.

19. Vercruysse J, Southgate VR & Rollinson D: The epidemiology of human and animal schistosomiasis in the Senegal River Basin. Acta Tropica, 42:249-259, 1985.

20. Vercruysse J, Fransen J, Southgate VR & Rollinson D: Pathology of _Schistosoma curassoni_ infection in sheep. Parasitology, 91: 291-300, 1985.

21. Walker TK, Rollinson S & Simpson AJG: Differentiation of _Schistosoma haematobium_ from related species using cloned ribosomal RNA gene probes Molecular and Biochemical Parasitology, 20: 123-131, 1986.

OCCURENCE OF HUMAN LUNG FLUKE INFECTION IN AN ENDEMIC
AREA IN LIBERIA

R. SACHS

1. ABSTRACT
    Attention was focussed on human paragonimiasis in Liberia
during the last years. Many cases will go unnoticed unless
physicians have an eye for them and know to differentiate
lung fluke infection from pulmonary tuberculosis. Therefore,
prevalence surveys in schools and villages are needed to
assess the true extent in an endemic area.

Of 25 patients examined for the presence of lung fluke eggs
at the Bong Mine Hospital in 1984 and 1985, 7 paragonimiasis
cases were found. Earlier examinations (Jan/Feb 1983) of
stool and sputum samples of 127 children had revealed 9
cases. Adding 8 further incidental findings since 1981 brings
the latest total up to 24 cases of paragonimiasis within the
Mawua/Haindi focus. Size and morphology of the eggs clearly
identified them as those of Paragonimus uterobilateralis.
The investigations show that lung fluke infection is more
common in Liberia than has been assumed. Close cooperation
between the public health authorities is needed to control
this food-borne zoonosis.

2. INTRODUCTION
    In many developing countries the population comes into con-
tact with disease organisms whilst in search of the elemen-
tary needs of life, such as water and food. Lung fluke infection of man is one of those diseases closely connected to
water and food, since freshwater crabs from local creeks and
rivers constitute an important supply of animal protein in
many tropical rain forest zones.

Human infection is acquired from ingesting raw or insufficiently cooked crab material containing live paragonimus metacercariae. Epidemiologically, the parasite shows little
host-specifity and can develop in a great variety of mammals,
including man. Compared with other tropical diseases and considering frequency and clinical symptoms, the morbidity of
African paragonimiasis appears to be insignificant. The life-
cycle of the parasite occurs principally in the African civet Viverra civetta (7), the long-nosed mongoose Crossarchus
obscurus (8) and other members of the viverrid family (3)
which act as final hosts, and snails and freshwater crabs as
intermediate hosts. Domestic animals are only accidentally
infected, and human beings also get involved only when

accidentally intruding the natural cycle by eating infected crabs. Thus paragonimiasis is a true food-borne zoonosis and of mutual interest for both the veterinary and the medical professions.

3. DIFFERENTIAL DIAGNOSIS AND SITUATION IN LIBERIA

Pulmonary paragonimiasis is often confused with tuberculosis since the clinical picture, viz. chronic cough, hemoptysis and the radiographic demonstration of chest shadows is strongly parallel in both infections. It is the well-known history of zoonoses, that they often are only acknowledged in an area when the first human case is reported. Presence of paragonimus metacercariae in the freshwater crabs Liberonautes latidactylus was demonstrated already in the late 60's (6), but since no human cases had been known to occur in Liberia at that time, the observation was neglected as being of no medical importance. 10 years later, demonstration of paragonimus eggs in human sputum as the causative organisms of hemoptysis brought about considerable change in focus with new emphasis on paragonimiasis. For the first time an explanation could be given for those hemoptic patients who did not respond to tuberculosis treatment.

Our investigations (5) at local hospitals revealed that the first cases of paragonimiasis in Liberia had been recorded in 1979 (1 case: at ELWA-hospital) and in 1980 (4 cases: at ELWA-, Zorzor-, and Buchanan-hospitals). The operculated, golden-brownish paragonimus eggs in the sputum of these 5 patients were errouneously entered in the concerned hospital's laboratory reports as ova of the East Asian lung fluke Paragonimus westermani. Such a diagnosis is very doubtful from zoo-geographical viewpoints. When we considered an existing endemicity of human paragonimiasis in Liberia itself, it was argued against us that the infections may had been acquired outside Liberia. However, the measurements and morphology of the paragonimus eggs excreted by the Liberian patients and examined by us as from 1981 onwards clearly showed that not the Asian lung fluke but the indigenous Paragonimus uterobilateralis was the causative organism in Liberia.

In close cooperation with the laboratory staff of government and missionary hospitals, the presence of 3 endemic foci was established: two in northern and southern Lofa County (in the Popalahun and Zorzor areas) and one in Bong County (in the Mawua/Haindi area). This latter focus is subject to this paper.

4. THE MAWUA/HAINDI FOCUS

The first cases of human infection in this endemic area came to our knowledge when, on 17.12.1981, we found paragonimus eggs in the sputum of a 6 years old girl from Mawua, a small settlement on the St. Paul river 5 km west of the district village Haindi (4). As the majority of the previously recorded paragonimus patients were also children,

case-finding activities were targeted to the young and
adolescent group. In January and February 1983, stool and
sputum samples of all Mawua-children up to 14 years of age
were collected and examined for paragonimus ova. From a total
of 127 children 5 girls aged 3-7 years and 4 boys aged 5-14
years were positive (2). None of the patients or their
parents were aware of the infections. The "rusty colour" of
the sputum observed from time to time, was attributed to the
red laterite dust of the African roads. During school-visits
and discussing with the pupils the common presenting symptoms
and favouring factors for the spread of the disease, such
discussions lead to the finding of other paragonimiasis-
infected household associates. This was the case during my
1985-survey in the Haindi-Government-School (unpublished,
now included in Table 1). After the examination of 70 school
children, an epileptic child who did not attend school was
brought, and lung fluke eggs were demonstrated in sputum and
stool.

Year	Programme	Number examined	Cases found	Age group
1981/82	Incidental findings	–	4	children
1983	Prevalence study/village	127	9	children
1984	Hospital/Lab records	13	5	not mentioned
1983/85	Incidental findings	–	3	children
1985	Hospital/Lab records	12	2	not mentioned
1985	Prevalence study/school	70	1	child

Table 1: Confirmed human lung fluke infections in the endemic
Mawua/Haindi focus, Bong County, Liberia, in the
4-year-period 1981-1985.

All incidental findings and results of prevalence surveys as
well as the Hospital Laboratory records of paragonimiasis
cases diagnosed within the Mawua/Haindi focus in the period
1981-1985 are summarized in Table 1. Since the infections
were found in the close vicinity of the Bong Mine Hospital,
all patients reporting there with a history of hemoptysis
are now subjected, in addition to the routine examination for
acid fast bacilli (AFB), to a special sputum examination for
paragonimus eggs. In 1984, of 13 patients thus examined 2
were AFB-positive, and 5 sputum samples contained paragonimus
eggs. In 1985 2 new cases of paragonimiasis were recorded at
the Bong Hospital Laboratory (Table 1): of 7 patients examined
for both AFB and lung fluke eggs, 1 patient was AFB-negative
but paragonimus-egg-positive whereas the second patient was
suffering from both tuberculosis and paragonimiasis. This
finding stresses how important it is for the clinician and
the public health authorities alike, to bear paragonimiasis
in mind in suspect tuberculosis patients.

5. DISCUSSION

It is evident that cases of human paragonimiasis are more common in endemic areas than the number of reported cases indicates. Prevalence studies in schools and villages, and the close cooperation with hospital laboratories are more likely to show the true extent of the disease. From the medical point of view, the introduction of Praziquantel as a very effective remedy against <u>Paragonimus uterobilateralis</u> in Liberia (1) is a good tool to control human infections. From the public health point of view, however, treatment of the infected patients alone - as efficient the drug may be - does not at all suffice because the animal reservoir and the actual source of infection remain untouched. The wild animals serving as a natural reservoir can hardly be controlled, and it is virtually impossible - and it can not be our aim - to eliminate the freshwater crabs in the area. New human infections can only be prevented by teaching the population to improve the standard of food preparation and by them avoiding eating raw crab material. If the goal "Health for All by the Year 2000" is to be attained, then it is time to start with more enlightenment of the rural people on disease causing organisms in general, and the means and methods to prevent food-borne diseases in particular.

REFERENCES

1. Monson MH, Koenig JW and Sachs R: Successful treatment with Praziquantel of six patients infected with the African lung fluke Paragonimus uterobilateralis. American Journal of Tropical Medicine and Hygiene, 32: 371-375, 1983.
2. Sachs R, Albiez EJ and Voelker J: Prevalence of Paragonimus uterobilateralis infection in children in a Liberian village. Transactions of the Royal Society of Tropical Medicine and Hygiene, 80: (in press)
3. Sachs R and Kern P: Epidemiological investigations of human lung fluke infection in Gabon, Central Africa. 8th International Congress Infectious and Parasitic Diseases, Stockholm, Sweden; Abstr. p. 121, 1982.
4. Sachs R and Voelker J: Human paragonimiasis caused by Paragonimus uterobilateralis in Liberia and Guinea. Tropenmedizin und Parasitologie, 30: 15-16, 1982.
5. Sachs R and Voelker J: Lungenegelinfektion beim Menschen in Liberia. In: Medizin in Entwicklungsländern Bd. 16: Tropenmedizin, Parasitologie: Boch J(ed) 455-458, Verlag Peter Lang, Frankfurt/Bern / New York, 1984.
6. Voelker J: Morphologisch-taxionomische Untersuchungen über Paragonimus uterobilateralis (Trematoda, Troglotrematidae) sowie Beobachtungen über den Lebenszyklus und die Verbreitung des Parasiten in Liberia. Zeitschrift für Tropenmedizin und Parasitologie, 24: 4-20, 1973.

7. Voelker J and Sachs R: Observations on the life-history of Paragonimus uterobilateralis in West Africa: the African civet (Viverra civetta) as natural reservoir in Nigeria. Proceedings 3rd International Congress of Parasitology, München, 529-530, 1974.
8. Vogel H and Crewe W: Beobachtungen über die Lungenegel-Infektion in Kamerun (Westafrika). Zeitschrift für Tropen-medizin und Parasitologie, 16: 109-124, 1965.

LARVA MIGRANS IN PERSPECTIVE

E.J.L. SOULSBY

1. INTRODUCTION
  Larva migrans, visceral or cutaneous, may be caused by a variety of
ascarids, for example, Toxocara, Toxascaris, Baylisascaris,
Lagochilascaris, Ascaris spp., as well as Capillaria hepatica in the
case of the visceral form and various hookworms, but principally
Ancylostoma braziliensis, with respect to cutaneous larva migrans.
However, the parasite of major concern is Toxocara canis of the dog and
its role in visceral larva migrans (VLM) and ocular larva migrans (OLM)
of man.
  Toxocara canis is a ubiquitous ascarid of dogs and various studies
have concluded that 20-90% of puppies are infected with the parasite.
The global population of dogs is not known, but it is estimated that
there are 60 to 80 million in the USA, 6.3 million in the UK (in five
million households) and comparable numbers in various countries of
western Europe. Accurate figures for dog populations are not available
for the majority of countries and in some the number of dogs has been
reduced drastically for disease control reasons (e.g. rabies, hydatid).
However, it is possible that 200-300 million dogs owned by approximately
the same number of households is an underestimate of the global reservoir of Toxocara infection of dogs for humans. The question arises
whether or not this "sea of Toxocara around us" is translated to human
exposure to the parasite.

2. SOURCES OF TOXOCARA INFECTION FOR DOGS
  Eggs of T. canis are subglobular with a thick, finely pitted shell, 75
by 90 microns. Embryonation to the infective stage (eggs containing a
second stage larva or possibly third stage according to Araujo (3))
occurs in 10 to 15 days under optimal conditions (27°C). On ingestion
the eggs hatch in the duodenum, and the subsequent life cycle of T.
canis is complex. There are several routes by which dogs become
infected with adult worms. Infection in the dog is characterised by
somatic migration of second stage larvae to the tissues, muscles being
the predominant site. Such larvae are long-lived, perhaps for three to
10 years, and their activation in the reproductive bitch leads to mater-
nal transfer of infection. Larvae become activated about the 42nd day
of pregnancy, migrate across the placenta to the fetal liver and, at
birth, migrate to the lungs where they pass to the alveoli and thence
bronchioles and trachea and are swallowed to mature in the small
intestine in three to four weeks after birth - though eggs may be passed
by precocious developing worms as early as 14 days after birth.
Transplacental infection predominates in bitches which have acquired
dormant second stage larvae prior to pregnancy or within the first 30-45

days of pregnancy. In utero infection occurs in 98% of experimental infections (7), but it is not known whether this also applies to natural infections. Transmission may also occur by the mammary gland and larvae occur in the milk for the first three to four weeks of lactation, peaking during the second week of lactation. Transmission through the milk can occur when bitches have been infected prior to pregnancy, though the number of larvae transmitted in this manner is low (1.5%) (7). However, when bitches are infected in late pregnancy or during lactation the galactogenic transfer of infection is an important source of infection for the nursing puppy (31,55). Larvae acquired via the milk mature in three to four weeks but some may produce eggs within two to three weeks of the puppies' birth.

As well as the transplacental and transmammary routes of infection, adult worm infections of puppies may result from the ingestion of infective eggs before the puppies reach three months of age. Such infections undergo a tracheal migration, eggs being produced by adult worms in 35 days after infection. The ability to undergo tracheal migration decreases markedly as the puppy ages and after three months the majority of newly acquired infective larvae are distributed in the somatic tissues.

Finally, dogs may be infected through the ingestion of paratenic hosts (e.g. rodents, chickens) infected with larvae of T. canis (23,58). Larvae acquired in this manner mature without further migration to become adult worms. Also, infections with mature T. canis are common and may be heavy in the lactating bitch primarily through her ingestion of immature stages swept out with the faeces of her pups (29,53).

Hence the life cycle of the parasite, with its variations, focusses adult infection in the puppy and in the lactating bitch. Various studies have demonstrated that dogs below six months of age are substantially infected (e.g. 30-40%) whereas animals over six months are lightly infected (e.g. 4-9%) (23). Egg production by puppies and lactating bitches may reach 1.5×10^7 eggs per day (29) and these may accumulate and survive in the soil for several months or years, hence there is every opportunity for infection to occur in the dog.

The life cycle of T. cati in the cat is similar to that of T. canis except that tranplacental transfer of larvae does not occur. Hatched larvae migrate in the tracheal route producing adult infections in about 50 days, or encyst in somatic tissues. Transmammary infection is common and probably the majority of kitten infections are derived from the milk of infected queens (51). Furthermore, paratenic hosts play an important role in T. cati infection, earthworms, cockroaches, birds and rodents serving as such.

3. SOURCES OF TOXOCARA INFECTION FOR HUMANS

Humans become infected by the accidental ingestion of embryonated eggs, through contamination by infected soil of food, fingers, toys, etc. Infective eggs behave in man as they do in other paratenic hosts and second stage larvae are distributed to various tissues, especially liver, lungs, eye and brain.

Highest concentrations of T. canis eggs occur in kennels housing numbers of dogs, breeding kennels and dog training schools being particularly heavily contaminated (23). The environmental contamination resulting from companion urban and suburban dogs and stray dogs varies considerably. For example, contamination of public parks in Perth,

Australia, was not detected in 66 soil samples from these or from 200 sand samples of "dog beaches", that is beaches on which dogs are allowed to exercise and which become quite heavily contaminated by faeces (15). On the other hand 18-23% of soil samples were positive for T. canis eggs in highway rest stops, parks and playgrounds in Kansas (10). A 5.2% infection rate of soil samples collected from public parks was noted in the London area with a similar percentage for private gardens (37). In a survey of public parks in Glasgow, Scotland, 12% of soil samples contained viable eggs of Toxocara (41). Studies in Germany, Czechoslovakia, Scotland, the Sudan and Iraq showed similar contamination rates of soil in public parks and elsewhere (63,64).

However, dog ownership is not necessarily required for contamination of private gardens since Smith et al (49) reported that infected soil samples were equally common in gardens in Baton Rouge, Louisiana where a dog was not kept as where one was.

Egg survival depends on soil type and climatic conditions. Eggs are relatively short-lived in hot dry sandy soils (15) but may survive one to two years in temperate climates (23). Clay soils facilitate egg survival and concentration, eggs settling out below the surface being protected from sunlight and drying (5). The eggs are not destroyed by conventional sewage treatments and T. canis eggs may be found in sewage and effluents (17).

Exposure to contaminated environments alone appears not sufficient to produce infection or disease in humans. For example, individuals such as veterinarians and kennel workers commonly exposed to parasitised animals showed no greater risk of infection than matched controls (19). Similarly, employees in a greyhound racing kennel showed no greater seropositiveness than the general population (24). Nevertheless, a relationship between antibody titre to Toxocara and the length of time engaged in dog breeding has been shown (62) in that a significantly higher prevalence (15.7 per cent) of antibody titres occurred in 102 dog breeders than the 2.6 per cent of normal healthy adults in the UK. An increase of seroprevalence appeared to occur as the number of bitches whelped per year and the length of time engaged in dog breeding increased.

Nevertheless, in the UK, approximately 50% of patients with clinical toxocariasis have never owned a dog or cat or had close contact with pets(62). This has led to the conclusion that most human infections derive from contaminated soil in public places.

This contrasts with the situation in the USA where there appears to be a significant association between visceral larva migrans in children and the household dog as a source of infection (20). Furthermore, a familial clustering of clinical cases of toxocariasis has been reported (22) which could imply infection being derived from a domestic situation. In the case of ocular larva migrans a strong correlation occurred with the presence of puppies in the household (48).

Recent studies of the influence of host factors, environment and behaviour using 8,457 serum samples undertaken by the Centre for Disease Control, Atlanta, Georgia and other laboratories, showed an overall seroprevalence of 2.8% for Toxocara infection in the USA with the South and North East having the highest seroprevalences (21). Seroprevalence was higher for blacks than whites and overall it was highest in the age groups one to five and six to 11 in both blacks and whites. The possible relationship of these data to the practice of pica in a population is discussed by Glickman and Schantz (21) who suggest a causal relationship between pica and Toxocara infection.

Higher prevalences of seropositive reactions to T. canis are found in certain groups of individuals suggesting a greater exposure to infection or an involvement of the parasite in particular diseases. For example, compared with the 2% seropositive status of apparently normal healthy individuals, those with choroidoretinitis, uveitis, hepatomegaly and asthma showed 2, 10, 29 and 17% seropositivity, respectively (60). It has also been suggested (61) that human T. canis infection is associated with neurological disorders such as epilepsy and that either toxocaral infections can cause epilepsy or epileptics are more prone to the infection. More recent studies (21) have shown that epileptics do have a higher Toxocara antibody titre, however no differences in antibody titres were noted between children with idiopathic epilepsy and epilepsy of known origin and since pica, mental retardation and hyperactivity are associated with epilepsy, it was postulated that epilepsy predisposed to toxocariasis.

Whereas it is generally believed that human toxocariasis results from the ingestion of infective eggs of the parasite, little consideration has been given to paratenic hosts as a source of human infection. Brain and spinal cord lesions in pigs due to T. canis larvae have been reported (13) and larvae of T. cati migrate in pigs (46). A serological survey in the United Kingdom would suggest that pigs are not infrequently exposed to T. canis (54) and sheep may show hepatic lesions which are ascribed to Ascaris suum infection (32) but it is possible that similar lesions may be derived from T. canis infection of sheepdogs. It is known that T. canis larvae can migrate and survive in the tissues of many hosts (52) and hence may be a source of human infection should infected tissues be eaten in a raw or undercooked state. Indeed Beaver (4) suspected that toxocariasis was the cause of hypereosinophilia in individuals eating raw liver for the treatment of pernicious anaemia.

4. VISCERAL LARVA MIGRANS

There are significant differences in the relationships between visceral larva migrans (VLM) and ocular larva migrans (OLM). These are summarized by Glickman and Schantz (21) in that the age of the patient at diagnosis for VLM is young, less than 10 years, for OLM, two years to adult, and for diagnosis of combined VLM and OLM less than five years. The theoretical dose of T. canis eggs to induce these various manifestations is high, low and very high, respectively, and the incubation periods are short (days to months), long (months to years) and very short (days), respectively. It is likely that these are oversimplifications but nevertheless they are useful broad divisions of the clinical situation.

VLM patients are usually young, one to three years of age, they invariably have a history of geophagia and in the USA a significant relationship occurs between VLM and the presence of a household dog, particularly a puppy (48). As noted above, this relationship between VLM and a household dog is not as clear in the UK. VLM is characterised by eosinophilic granulomas in the liver and lungs, frequently detectable by biopsy. There is a persistent eosinophilia, hepatomegaly, hypergammaglobulinaemia, leukocytosis and pica (33). Pulmonary involvement is common but rarely fatal and transient pulmonary infiltrates occur on radiographs. The clinical and pathological picture is one consistent with an invasion by substantial numbers of larvae of T. canis and in fact up to 300 larvae per gramme of tissue has been recorded from a patient (4).

Patients with VLM tend to have higher ELISA titres for Toxocara than those with OLM (21) which is consistent with the concept that higher doses of infective eggs induce higher levels of antibodies. Multiple infections of various species, including primates, produced high antibody titres and a greater proportion of larvae are found in the liver than with single or lower doses of infective eggs (1). When high doses of eggs are administered to animals ocular lesions are produced (36) and this fits the concept that where VLM and OLM occur together, humans have been exposed at an early age to large doses of eggs.

Dodge (12) has hypothesised that in children who ingest large numbers of eggs, larvae are destroyed in the liver. However, in infections with a low number of eggs the immune response fails to destroy larvae which then continue to grow and migrate elsewhere in the body, including the retina, where they may be filtered out of the blood stream because of their size.

The principal lesion of VLM is an eosinophilic granuloma and larvae appear to be destroyed as early as 14 days in primates (2). The granuloma is T-cell independent in the initial stages but later its maintenance is T-cell dependent (56). The lesion has been considered to be immune-complex mediated, with a chemotactic factor being released from complement activated by immune complexes in the granuloma (35,57). Beaver (4) demonstrated motile larvae, infective for mice, in the livers of monkeys 10 years after infection. Larvae of T. canis have been shown to shed attached eosinophils in vitro (44) and perhaps this also occurs in vivo through the modulation of surface antigens since the turn-over of these has been shown to be rapid (30).

Diagnosis of VLM is based on an assessment of non specific evidence and specific serodiagnosis. VLM should be suspected when there is significant hepatomegaly, leukocytosis with a white blood count greater than $10,000/mm^3$, eosinophilia greater than 10%, an anti-A isohaemagglutinin titre of 1:400 or more and an anti-B titre of 1:200 or more, and IgM and IgG levels greater than two standard deviations above normal for age and gender.

An ELISA test using excretory and secretory products of in vitro maintained second stage larvae of T. canis is currently accepted as the most sensitive and specific test available for diagnosis. Isohaemagglutinins may complicate the test. Modification of the ELISA to detect the class of immunoglobulin may have important advantages. For example VLM patients show specific IgG and IgM antibodies whereas IgM antibodies are not usually detected in OLM patients (18). Further development of the ELISA to detect IgE specific antibodies may offer additional advantages.

5. OCULAR LARVA MIGRANS

The relationship between VLM and OLM is unclear: both are associated with the migration of T. canis larvae into host tissues but they appear to be separate clinical entities. However, Girdwood (18) cautions that the division of human toxocariasis into visceral and ocular, although clinically convenient, may be an oversimplification. OLM is usually seen in an older age group of children and even in adults but usually it occurs without current signs of VLM. OLM has been reported some years following an episode of VLM and occasionally OLM has been seen concurrently with VLM (50).

The concept that OLM is dose related has received emphasis recently. It is hypothesised that low doses of larvae provide insufficient antigen to stimulate a marked eosinophil and antibody level and consequently larvae migrate unimpeded eventually to be filtered out at random in a capillary which may be that of the retina or choroid (21). Larvae may survive in tissues for several years as noted above and this would be consistent with the long incubation period that appears to be part of the OLM condition.

The age prevalence for OLM is higher (mean 8-12 years) than VLM and pica is not a prominent association. Eosinophilia which is pronounced and persistent in VLM is usually absent with OLM. Detailed epidemiological studies of OLM show a significant association with dogs, especially pups in the home within one year prior to onset of ocular signs (47,48). In a study in Atlanta, Georgia, a significant association was found between OLM and the presence of puppies less than three months of age in the household within one year of clinical OLM (48). Whereas dog or puppy ownership is correlated with OLM in the USA in the UK this may not be the case since 50% of patients with clinical toxocariasis have never owned a dog or had close contact with one (62). Girdwood (18) reports an incidence in the UK of diagnosed Toxocara-induced disease in the order of two per million. In Scotland there are about ten cases per year and almost all are the ocular form.

Despite the apparent dissociation between VLM and OLM, previous experience of toxocariasis in an unapparent form may be necessary for the pathological manifestations of OLM. Schantz et al (47) have suggested that ocular manifestations are more likely to occur in previously infected individuals. Furthermore, de Savigny (11) induced retinal lesions and endophthalmitis only in animals which had been previously sensitized by infection when he inoculated hatched larvae into the ophthalmic branch of the internal carotid artery. Local antibody production was detected in the aqueous humour of affected eyes and not in the contralateral unaffected eyes. It has been stated that such findings are difficult to reconcile with the fact that many OLM patients are seronegative. However, IgE mediated reactions may be occurring in actively immunised hosts even when no IgE is detectable in serum (43). Further IgE specific antibodies may not be detected by the ELISA techniques at present in use. An additional point is that local ocular IgE production may occur in the absence of circulating IgE as shown by the presence of aqueous humour IgE antibodies in eyes infected with Ascaris or Toxocara larvae (45). On the other hand substantial levels of IgE in the serum may have a limited role in ocular pathology because blocking antibodies (e.g. IgG with the same specificity) may reduce the IgE mediated mast cell degranulation in vivo (45).

The pathogenesis of OLM has been studied in guinea pigs following the intravitreal introduction of second stage larvae of T. canis (42,43). These attract an eosinophil rich infiltrate within 24 hours which persists for 50 days. Larvae are surrounded by a granuloma of eosinophils by the 12th day and degranulation of eosinophils occurs. Also there is dense infiltration of the choroid with eosinophils alone, not immediately adjacent to a larva, and this is accompanied by damage to the overlying retina. These responses are more severe in animals which have been systemically pre-sensitized to ascarid antigens.

A significant decrease in mast cell numbers in infected eyes correlates well with an early IgE response and therefore may be associated with IgE-antigen induced degranulation of mast cells (43). This along with the degranulation of eosinophils could further damage ocular tissues.

Dead ascarid larvae evoked only a minor inflammatory reaction. Ascarid antigens can be detected in the aqueous humour by reversed Prausnitz-Kustner (PK) tests for up to 189 days after intravitreal inoculation suggesting that larvae may survive for some considerable time in the eye (14).

Diagnosis of OLM utilises the same tests as those for VLM. The ELISA at a titre of 1:32 has a sensitivity and specificity of 73 and 95 per cent respectively (38). The possible development of an ELISA to detect immunoglobulin class specific antibodies may advance diagnosis since IgM antibodies are not usually detectable in OLM patients (18). However, since 10% of OLM cases do not have a positive ELISA titre and also have a low peripheral eosinophilia (38), examination of vitreous or acquous humour may be useful and various authors have reported higher IgG ELISA titres in the vitreous or aqueous humour than in the serum of patients with OLM (28). Also, antigen of T. canis larvae has been detected in the eyes of guinea pigs injected intravitreally with larvae and the detection of antigen may be an additional diagnostic aid in man.

5. TREATMENT OF HUMAN TOXOCARIASIS

Anthelmintic treatment of VLM or OLM is generally unsatisfactory. Many cases are treated symptomatically with steroids to control inflammatory responses and, where anthelmintics are required, thiabendazole, albendazole and diethylcarbamazine have been used, usually in conjunction with steroids.

6. TREATMENT AND CONTROL OF CANINE TOXOCARIASIS

The reduction of environmental contamination is an essential factor in the control of the human infection. This can be achieved by the regular anthelmintic treatment of infected dogs.

A number of effective anthelmintics exist for patent T. canis infections in puppies and dogs. To control infection in puppies treatment should be given at 2, 4, 6, 8 and 12 weeks of age. This will prevent contamination of the environment with eggs from adult parasites derived from prenatal, lactogenic or direct oral egg infection in puppies (26,27).

The nursing bitch is of particular importance in the contamination of the environment. She passes eggs from parasites originating from larvae shed in the faeces of the puppy and also from her own parasites. She also is immunosuppressed at this time and this facilitates the acquisition and maturation of adult worms (29).

Treatment of bitches to reduce migrating or dormant larvae is now feasible. Fenbendazole given at a dose rate of 50mg/kg from 40 to 50 days of pregnancy until the end of the third week of lactation, killed migrating larvae and prevented transmission to puppies (6,16). However, if infections are at a high level in bitches and subsequently in puppies, fenbendazole treatment, even prolonged, will not eliminate transmission to the pups who may require subsequent treatment to eliminate adult parasites (8,26).

6. TOXOCARA VITULORUM AS A CAUSE OF HUMAN VLM

Toxocara vitulorum is a common ascarid of cattle, zebu and buffalo in many parts of the tropical world. The life cycle is very similar to that of T. cati in that infection of calves occurs neonatally through the milk. There is no evidence that transplacental infection occurs and it has not been possible to create adult worm infections in calves by feeding embryonated eggs.

Egg output of infected calves may reach 100,000 eggs per gram of faeces. Eggs are resistant to adverse conditions and may persist in the environment as do other ascarid eggs. It is possible, therefore, that heavy contamination of soil with T. vitulorum eggs occurs. However, there are no reports to provide an assessment of the situation in man. Larvae of T. vitulorum are able to migrate in other hosts and do so in mice to be distributed to various tissues. Such larvae survive only for a few weeks (59) which might suggest that human infection, should it occur, might be self limiting. The use of purified antigens for sero-diagnostic tests and monoclonal antibodies for the detection of infection in humans may be of use.

7. TOXOCARA PTEROPODIS AS A CAUSE OF HUMAN VLM

Toxocara pteropodis is an ascarid of the flying fox or fruit bat Pteropus poliocephalus and it has been incriminated as a cause of human VLM in the aboriginal community at Palm Island in North Queensland (9,34). Eggs shed by nursing fruit bats may contaminate fruit destined for human consumption. Larvae accumulate in the liver of adult fruit bats and probably remain viable there for years. In female bats larvae are mobilized during lactation and pass to the suckling bat in the milk (39).

Studies of the behaviour of larvae of T. pteropodis in mice indicated that larvae rapidly accumulated in the liver where they grew over a period of five to seven weeks. Neither pregnancy nor lactation stimulated further migration in mice (40). The hepatotropism demonstrable in mice is reflected in the clinical manifestation in humans, where hepatitis is a major component of the clinical disease in children (9).

8. BAYLISASCARIS PROCYONIS AS A CAUSE OF HUMAN VLM

Baylisascaris procyonis is a common ascarid of the racoon (Procyon lator) and has been recognised as the cause of often fatal larva migrans in animals, but recently it has been incriminated as the cause of two fatalities in children and also of human OLM. Human infection is associated with close association with racoons. In one case a pet racoon had been acquired a few weeks before symptoms of OLM developed and, in two fatalities associated with CNS lesions, one was associated with racoon faeces on an open fireplace contaminated by animals in the chimneys and in the other a child had chewed pieces of bark contaminated by eggs of the parasite.

Larvae of B. procyonis undergo somatic migrations similar to other ascarids but they also grow during migration (25) reaching 1.5 to 2 mm in length. They actively invade tissue such as the brain, eyes and somatic tissues and may also encyst in various tissues.

The prevalence of B. procyonis infection in racoons is unclear but in the mid western USA 70% of adult racoons and more than 90% of juveniles may be infected. The hazard to human health by this ascarid has yet to be fully assessed but obviously the keeping of racoons as pets constitutes a substantial risk. Straw, hay and earth in barns contaminated by racoon faeces are potential danger areas for children.

9. CONCLUSION

Ascarid infections of domestic and synanthropic animals commonly occur and may be translated to human infection in the form of VLM or OLM. The factors that determine this require further investigation and more exquisitely specific tests and reagents may assist in this. However despite the sea of Toxocara around us only a relatively small proportion is translated to human responses and even less to clinical manifestations.

REFERENCES

1. Aljeboori TL & Ivey MH: Toxocara canis infection in baboons. American Journal of Tropical Medicine and Hygiene 19: 249-254, 1970.
2. Aljeboori TL, Stourt C & Ivey MH: Toxocara canis infections in baboons. Distribution of larvae and histopathologic responses. American Journal of Tropical Medicine and Hygiene 19: 815-820, 1970.
3. Araujo P: Observacoes pertinentes as primeiras ecdises de larvas de Ascaris lumbricoides, A. suum e Toxocara canis. Revista del Instituto medicina tropical Sao Paulo 14: 83-90, 1972.
4. Beaver PC: Zoonoses with relation to parasites of veterinary importance, In: Biology of Parasites, (Ed. Soulsby EJL). New York: Academic Press, pp. 215-226, 1966.
5. Beaver PC: Biology of soil transmitted helminths: the massive infection. Health Laboratory Sciences 12: 116-125, 1975.
6. Bosse M & Stoye M: Zur Wirkung verschiedener Benzimidazolcarbamate auf somatische Larven von Ancylostoma caninum ERCOLANI 1859 (Ancylostomatidae) und Toxocara canis WERNER 1782 (Anisakidae). II. Untersuchungen an der graviden Hunden. Zentralblatt fur Veterinarmedizin 28: 265-279, 1981.
7. Burke TM & Roberson EL: Prenatal and lactational transmission of Toxocara canis and Ancylostoma caninum: experimental infection of the bitch before pregnancy. International Journal for Parasitology 15: 71-75, 1985.
8. Burke TM & Roberson EL: Fenbendazole treatment of pregnant bitches to reduce prenatal and lactogenic infections of Toxocara canis and Ancylostoma caninum in pups. Journal American Veterinary Medical Association 183: 987-990, 1983.
9. Byth S: Palm Island mystery disease. Medical Journal of Australia 2: 40-42, 1980.
10. Dada BJD & Lindquist WD: Prevalence of Toxocara spp eggs in some public grounds and highway rest areas in Kansas. Journal of Helminthology 53: 145-146, 1979.
11. de Savigny D: Quoted by Ogilvie BM & de Savigny D: In: Immunology of Parasitic Infections, (Eds. Cohen S & Warren RS). Oxford: Blackwell, p. 726, 1982.
12. Dodge JS: Toxocara canis: the risks of infection. New Zealand Medical Journal 91: 24-26, 1980.
13. Done JT, Richardson MD & Gibson TE: Experimental visceral larva migrans in the pig. Research in Veterinary Science 1: 133-151, 1960.
14. Donnelly JJ, Rockey JH, Stromberg BE, et al: Immunopathology of ocular Toxocara canis and Ascaris suum infection in the guinea pig. Ascarid antigens detected in aqueous humor by a reversed P-K test.

Investigative Ophthalmology and Visual Science 18: (suppl.) 95, 1979.
15. Dunsmore JD, Thompson RCA & Bates IA: Prevalence and survival of Toxocara canis eggs in the urban environment of Perth. Veterinary Parasitology 16: 303-311, 1984.
16. Duwell D & Strasser H: Versuche zur Geburt Helminthen-frier Hundwelpen durch Fenbendazole - Behandlung. Deutsche tierarztliche Wochenschrift 85: 234-241, 1978.

17. Fitzgerald PR & Ashley RF: Differential survival of Ascaris ova in wastewater sludge. Journal of Water Pollution Control Federation, 1722-1724, 1977.
18. Girdwood RWA: Human toxocariasis. Journal of Small Animal Practice 27: 655-661, 1986.
19. Glickman LT & Cypess RH: Toxocara infection in animal hospital employees. American Journal of Public Health 67: 193-195, 1977.
20. Glickman LT, Chandry IU, Constantino J, et al: Pica patterns, toxocariasis and elevated blood lead in children. American Journal of Tropical Medicine and Hygiene 30: 77-80, 1981.
21. Glickman LT & Schantz PM: Epidemiology and pathogenesis of zoonotic toxocariasis. Epidemiological Reviews 3: 230-250, 1981.
22. Heiner DL & Kevy SV: Visceral larva migrans: report of the syndrome in three siblings. New England Journal of Medicine 254: 629, 1956.
23. Jacobs DE, Pegg EJ & Stevenson P: Helminths of British dogs: Toxocara canis - a veterinary perspective. Journal of Small Animal Practice 18: 79-92, 1977.
24. Jacobs DE, Woodruff AW, Shah AI, et al: Toxocara infections and kennel workers. British Medical Journal 1: 51, 1977.
25. Kazacos KR: Racoon ascarids as a cause of larva migrans. Parasitology Today 2: 253-255, 1986.
26. Lloyd SS: Toxocara canis: infection, control and treatment. Veterinary Annual 25: 368-375, 1985.
27. Lloyd SS: Toxocariasis. Journal of Small Animal Practice 27: 655-661, 1986.
28. Lloyd SS: Immunobiology of Toxocara canis and visceral larva migrans, In: Immune Responses in Parasitic Infections: Immunology, Immunopathology and Immunoprophylaxis. Vol. I: Nematodes. (Ed. Soulsby EJL). Boca Raton: CRC Press, 1986 In press.
29. Lloyd SS, Amerasinghe PH & Soulsby EJL: Periparturient immuno-suppression in the bitch and its influence on infection with Toxocara canis. Journal of Small Animal Practice 24: 237-247, 1983.
30. Maizels RM, de Savigny D & Ogilvie BM: Characterisation of surface and excretory-secretory antigens of Toxocara canis larvae. Parasite Immunology, 6: 23-37, 1984.
31. Manhardt J & Stoye M: Zum Verhalten der Larven von Toxocara canis Werner 1782 (Anisakidae) wahrend und nach der Lungenwanderung im definitiven Wirt (Beagle). Zentralblatt fur Veterinarmedizin 28: 386-406, 1981.
32. Mitchell GBB & Linklater KA: Condemnation of sheep livers due to ascariasis. Veterinary Record 107: 70, 1980.
33. Mok CH: Visceral larva migrans - a discussion based on review of the literature. Clinica Pediatrica 7: 565-573, 1968.
34. Moorhouse DE: Toxocariasis. A possible cause of Palm Island mystery disease. Medical Journal of Australia 1: 172-173, 1982.
35. Moreau AF, Mary R & Junod C: La toxocarose: malade allergique a

precipitins. Review of Allergy 11: 271, 1971.

36. Olson LJ: Ocular toxocariasis in mice: distribution of larvae and lesions. International Journal for Parasitology, 6, 247-251, 1976.

37. Pegg EJ: Dog roundworms and public health. Veterinary Record 97: 78, 1975.

38. Pollard ZF, Jarrett NH, Hagler WS, et al: ELISA for diagnosis of ocular toxocariasis. Ophthalmology 86: 743-749, 1979.

39. Prociv P: Observations on the transmission and development of Toxocara pteropodis (Ascaridoidea, nematoda) in the Australia Grey-headed Flying-Fox Pteropus poliocephalus (Pteropodidae: Megachiroptera). Zeitschrift fur Parasitenkunde 69: 773-781, 1983.

40. Prociv P: Observations on Toxocara pteropodis infections in mice. Journal of Helminthology 59: 267-275, 1985.

41. Quinn R, Smith HV, Girdwood RWA, et al: Studies on the incidence of Toxocara and Toxascaris spp. ova in the environment. I. A comparison of flotation procedures for recovering Toxocara spp. ova from soil. Journal of Hygiene 84: 83-89, 1980.

42. Rockey JH, Donnelly JJ, Stromberg BE, et al: Immunopathology of Toxocara canis and Ascaris suum infections of the eye: the role of the eosinophil. Investigative Ophthalmology and Visual Science, 18: 1172-1184, 1979.

43. Rockey JH, Donnelly JJ, Stromberg BE, et al: Immunopathology of ascarid infection of the eye. Role of IgE antibodies and mast cells. Archives of Ophthalmology 99: 1831-1840, 1981.

44. Rockey JH, John T, Donnelly JJ, et al: In vitro interaction of eosinophils from ascarid-infected eyes with A. suum and T. canis larvae. Investigative Ophthalmology and Visual Science, 24, 1346-1351, 1983.

45. Rockey JH, Donnelly JJ, John T, et al: IgE antibodies in ocular immunopathology. In: Advances in Immunology and Immunopathology of the Eye, (Eds. O'Connor GR & Chandler JW). New York: Maison Publishing pp. 199-202, 1985.

46. Roneus O: Parasitic liver lesions in swine experimentally produced by visceral larva migrans of Toxocara cati. Acta Veterinaria Scandinavica 4: 170-186, 1963.

47. Schantz PM, Meyer D & Glickman LT: Clinical, serologic and epidemiological characteristics of ocular toxocariasis. American Journal of Tropical Medicine and Hygiene 28: 24-28, 1979.

48. Schantz PM, Weiss PE, Pollard ZF, et al: Risk factors for toxocaral ocular larva migrans: a case control study. American Journal of Public Health 70: 1269-1272, 1980.

49. Smith RE, Hagstad HV & Beard GB: Visceral larva migrans: a risk assessment in Baton Rouge, Louisiana. International Journal of Zoonoses 11: 189-194, 1984.

50. Snyder CH: Visceral larva migrans. Pediatrics 28: 85-91, 1961.

51. Soulsby EJL: Helminths, Arthropods and Protozoa of Domesticated Animals. 7th edition. London, Bailliere Tindall, 1982.

52. Sprent JFA: The life history and development of Toxocara canis (Werner, 1782) in the dog. Parasitology 48: 184-209, 1958.

53. Sprent JFA: Post-parturient infection of the bitch with Toxocara canis. Journal of Parasitology 47: 284, 1961.

54. Stevenson P: Toxocara and Ascaris infection in British pigs: a serological survey. Veterinary Record 175: 1270-1273, 1979.

55. Stoye M: Galaktogene and Prenatale Infectionen mit Toxocara canis

148

beim Hunde (Beagle). Deutsche tierarztliche Wochenschrift 83: 107–108, 1976.

56. Sugane K, & Oshima T: Eosinophilia, granuloma formation and migratory behaviour of larvae in the congenitally athymic mouse infected with Toxocara canis. Parasite Immunology 4: 307–318, 1982.

57. Sugane K & Oshima T: Activation of complement in C-reactive protein positive sera by phosphorylcholine-bearing component isolated from parasite extract. Parasite Immunology 5: 385–395, 1983.

58. Warren EG: Infections of Toxocara canis in dogs fed infected mouse tissues. Parasitology 59: 837–841, 1969.

59. Warren EG: Observations on the migration and development of Toxocara vitulorum in natural and experimental hosts. International Journal for Parasitology 1: 85–99, 1971.

60. Woodruff AW: Toxocariasis: a public health problem. Environmental Health 1: 29.

61. Woodruff AW, Bisseru B & Bowe JC: Infection with animal helminths as a factor in causing poliomyelitis and epilepsy. British Medical Journal 1: 1576–1579, 1966.

62. Woodruff AW, de Savigny D & Jacobs DE: Study of toxocaral infection in dog breeders. British Medical Journal 2: 1747–1748, 1978.

63. Woodruff AW, Watson J, Shikara I, et al: Toxocara ova in the Mosul district, Iraq and their relevance to public health measures in the Middle East. Annals of Tropical Medicine & Parasitology 75: 555–557, 1981.

62. Woodruff AW, Salih SY, de Savigny D, et al: Toxocariasis in the Sudan. Annals of Tropical Medicine & Parasitology 75: 559–561, 1981.

IMMUNOLOGICAL STUDIES ON <u>ASCARIS SUUM</u> INFECTIONS IN MICE

PHAN VAN THAN and F.VAN KNAPEN

ABSTRACT
  <u>Ascaris lumbricoides</u>, the human roundworm is one of the most common intestinal parasites in the world. Prospects for the development of a dead vaccin against <u>Ascaris</u> infection were evaluated on the basis of a literature study and various experiments.
  An animal (mouse) model was worked out to study the larval migrating behaviour in different organs following experimental (oral) infection with embryonated eggs of <u>Ascaris suum</u>. In this study we report the use of ES antigen of L2 larvae from both <u>A.suum</u> and <u>A.lumbricoides</u> for immunisation studies. The influence of this immunisation upon numbers of migrating larvae after a challenge infection was studied. Initital studies were performed to characterise the antigens by immunoblotting techniques. In the mouse model larvae appeared in the liver between 3–5 days after infection and a peak number of larvae reached the lungs at day 7. These days were choosen to study a possible protective effect from immunisation with the ES antigens. The antigens caused moderate to good protection. ES antigens from <u>A.suum</u> also protect against challenge infections with <u>A.lumbricoides</u> in mice. In a Western Blotting technique ES antigens show limited numbers of antigenic bands with slight differences in molecular weight, depending on the developmental stage of the larvae.

1. INTRODUCTION
  Prospects for vaccins against the most widespread and serious nematode infections in man, Ascaris, are not very encouraging (4, 9). Vaccination against <u>Ascaris</u> with secreted larval antigen was reasonably succesful in some experimental animals (a.o. 5). In this study an attempt was made to test the suitability of ES antigens of L2 larvae from <u>Ascaris suum</u> to induce protective immunity in mice. For this purpose an experimental animal model to test the immunogenicity of the antigens was established before protection studies were carried out. Challenge infection was carried out with embryonated eggs of <u>Ascaris suum</u>. Enzyme Linked Immunosorbent Assay (ELISA) was used to follow the humoral responses. Cross protection was studied with <u>Ascaris lumbricoides</u> antigen in order to eliminate possible future problems with regard to availability of human material.

2. MATERIALS AND METHODS
2.1. <u>Mice</u>
  In all experiments with mice, random bred, conventional female Swiss mice were used of about 6 weeks of age. They were bred at the Institute.

2.2. Rabbits

New Sealand white rabbits ($\pm$ 2,5 kg) were obtained from the Institutes spf-colony. The rabbits were infected orally with 100.000 embryonated eggs and killed seven days later. The lungs were collected and processed in a Baermann technique to obtain migrating L3 larvae.

2.3. Parasites

Ascaris suum adult worms were collected in a pig abattoir in the Netherlands. Females were used to obtain eggs directly from the uteri. Washed eggs were deposited in petri dishes containing 0,5% formaldehyde and placed at room temperature for embryonation and maturation. Three weeks later the embryonated eggs were stored in flasks at $+4^{\circ}C$ until use.

Ascaris lumbricoides adult worms were obtained from human patients in a large hospital in Hanoi (Vietnam). Collection of eggs, embryonation and storage was done as mentioned above.

2.4. Hatching and collection of larvae

The embryonated eggs were washed at least three times in distilled water to remove formaldehyde. Thereafter the eggs were processed as described in detail by Adams Oaks and Kayes (1) to obtain sterile second stage larvae.

2.5. In vitro maintenance

The maintenance of both L2 and L3 larvae of either Ascaris species was carried out as described in detail earlier for Toxocara (7, 11). In preliminary studies it was found that L2 larvae die within a week in a medium without serum. Therefore the culture fluid of the L2 larvae was enriched with 400 ng of glycol-L-histidyl-L-lysine acetate tetrahydrate (Calbiochem) per 20 ml of culture medium (13). The actual maintenance of the larvae at $37^{\circ}C$ and collection of supernatant fluid was carried out for approximately 3 weeks or shorter if the presence of dead larvae became obvious. As soon as the culture medium colour turned from red into yellow, the supernatant fluid was collected and fresh medium was added. All supernatants (in vitro released antigens or ES-antigens) were pooled and stored at $-20^{\circ}C$ until use or further preparation.

2.6. Enzyme Linked Immunosorbent Assay (ELISA)

A sandwich ELISA was carried out with different antigens in order to demonstrate antibodies in the mouse sera. Somatic antigens prepared from the L2 and L3/4 larvae were diluted in sodium carbonate buffer (0.1 M, pH 9.6) until the protein content was 5 g/ml. Coating of the micro-ELISA plates was carried out in 100 µl quantities at $37^{\circ}C$ for µl hour. The ES antigens were coated by overnight incubation at $37^{\circ}C$ with undiluted culture fluid until complete evaporation had occurred. Before serial dilutions were added, the coated plate was washed twice with tapwater containing 0,05 % Tween 20 for 30 seconds each time in an automatic washing device (10). Serial dilutions were prepared starting at 1:10 with PBS (0.01 M, pH 7.2) containing 2 % bovine serum albumin (BSA) and 0.05 % Tween 20. Incubation was done at $37^{\circ}C$ for one hour. After washing (as above) the conjugate was added, optimally diluted in PBS and 0.05 % Tween 20. Conjugate was a peroxidase labeled rabbit-anti-mouse IgG (H + L) commercially obtained from Cappel Laboratories, USA. Incubation was done for 1 hour at $37^{\circ}C$. After washing again the substrate was added consisting of 80 mg 5-aminosalicylic acid (5-amino-2-hydroxy benzoic acid) in 100 ml of distilled water adjusted to a pH of 6.0. Then, 0.05% H_2O_2 was added to the substrate in a 1:9 (v/v) ratio. After incubation for one hour at roomtemperature, absorbance was measured with a Multiscan micro-ELISA photometer (at 449 nm) and polaroid photographs were taken for visual interpretation.

2.7. SDS-polyacrylamide gelelectrophoresis and immunoblotting

Antigen samples were electrophoretically separated by the method of Laemli (8). After preparation of the 50 µl antigen samples by boiling for 5 minutes with 10 µl of SDS (10 %), 10 µl of distilled water and 10 µl 2-mercapto ethanol, they were applied to the stacking gel. An electrophoretic run was made with 60-100 voltage using a constant current of 20 mA in the stocking gel followed by and 40 mA in the running gel. The total running time was approximately 3 hours until the bromiumphenolblue dye had reached the bottom. The separated proteins were transferred immediately to nitrocellulose sheets essentially as described by Towbin et al. (14). It was carried out in a Trans Blot Cell^R (BioRad), according to the manufacturers prescription. The transfer was made overnight with 0,1 KV and 10 mA. The staining procedure was carried out according to the Peroxidase anti-peroxidase method described by Glass et al. (6).

2.8. Protection Experiment

Cross immunisation of mice with ES-antigens of L2 larvae of Ascaris suum and Ascaris lumbricoides.

Antigen preparation and immunisation:

ES antigen (culture fluid) of L2 larvae was lyophilised and concentrated 3 times. 0.3 ml of antigen was thoroughly mixed with Freunds Complete Adjuvans (FCA) for subcutaneous injection at day 0 and with Freunds Incomplete Adjuvans (FIA) for booster injection at days 14 and 28. Control groups were injected with similar volumes (0.6 ml) of FCA and FIA alone at days 0, 14 and 28. At last control groups were not pretreated at all, but obtained oral challenge infection alone. Groups of 10 mice each were treated as follows:

1. A.suum/ES antigen immunised, challenged orally with
   A.suum eggs (2000).
2. A.suum/ES antigen immunised, challenged orally with
   2000 A.lumbricoides eggs.
3. A.lumbricoides/ES antigen immunised, challenged orally with
   2000 A.lumbricoides eggs.
4. FCA/FIA stimulated, challenged orally with
   2000 A.suum eggs.
5. FCA/FIA stimulated, challenged orally with
   2000 A.lumbricoides eggs.
6. No immunisation, infected orally with
   2000 A.suum eggs.
7. No immunisation, infected orally with
   2000 A.lumbricoides eggs.

Obduction was carried out with 5 mice of each group at day 45 for blood and liver collection. The remaining 5 mice per group were killed and bled at day 49 for blood and lung collection.

3. RESULTS AND DISCUSSION

The risk of inducing allergenic reactions due to the application of unpurified Ascaris antigens is a well-known phenomenon. In our experiment therefore ES antigen was tested as a possible candidate for immunisation against challenge infection in mice. Although ES antigens from the L2 stage of Ascaris suum could produce protective immunity in mice but not in guinea pigs (5, 12), the production of such antigen caused problems with regard to larval survival in culture medium. The introduction of glycyl-1-histidyl-1-lysine tetrahydrate (13) in the medium as a substitute for serum made longer in-vitro cultures possible. In fact our goal was to study protection against Ascaris lumbricoides. From a practical standpoint heterologous immunisation is preferable since large quantities of materials of human origin (A.lumbricoides) are needed.

Results of the cross-immunisation study between Ascaris suum and Ascaris lumbricoides with the help of ES/L2 antigens are presented in Tables 1 and 2 regarding the recovery of migrating larvae after challenge infection. Protection occurred in all immunised groups of mice both when larvae were counted in the livers (day 3 post challenge infection) and the lungs (day 7 post challenge infection). The reduction of migrating A.suum in the livers was 86 % in the homologous system (group I versus IV and VI). The reduction of A.lumbricoides larvae in the livers was 92 % in the homologous system (group III versus V and VII) and 90 % in the heterologous system (group II versus V and VII).

Similarly the reduction of migrating larvae of A.suum in the lungs was 84 % in the homologous system, whereas the reduction of A.lumbricoides larvae was 94 % in the homologous system, and 91 % in the heterologous system. Particularly in the lungs quite often no larvae at all were found (total protection) although ironically this phenomenon was also observed in a few control animals. This is the reason that statistically (Wilcoxon's test) no significant differences were observed.

With regard to the animal (mouse) model the following remarks can be made. In working out the animal model it was found that remarkable individual differences in recovery of larvae existed. In fact most larvae seem to be trapped already at the gut-level even in naive mice, since they were not found in any organ system besides liver (peak day 3) and lungs (peak day 7). (Data not shown here.) Although this is in accordance with results obtained by other investigators in literature (a.o. 2) even when other animal species were used like guinea pigs or rabbits this does not necessarily mean that migrating larvae behave similarly in immunised animals. This was not studied here nor in any study reported in literature. Therefore, the large reduction of migrating larvae even in the livers may suggest inhibition or trapping at a very early (intestinal) stage of the challenge infection but does not prove it. This was also suggested by Bindseil (3).

The relative lower counts for migrating larvae in immunised and control groups when challenge infections were given with A.lumbricoides eggs is an indication that the infectivity of the eggs, obtained from Vietnam may have been reduced because of the transport and older age. Another explanation could be that A.lumbricoides behaves differently in mice as compared to A.suum. This remains to be investigated. However, the ES antigens derived from L2 larvae from A.lumbricoides eggs did produce comparable immunity.

The results of the serological examination with the sera of the mice at the end of the experiment are summarised in Table 3. ES antigens of L2 A.suum are recognised by all immunised groups. The ES antigens of L2 A.lumbricoides are particularly recognised by the groups immunised with A.lumbricoides antigen. The ES antigen of L3/4 larvae of A.suum are poorly recognised by all groups, whereas the somatic antigens of either stage of larvae cross react serologically with all animals immunised. In fact, these data suggest a predominant immunological reaction for homologous antigens. This was also observed in the preliminary immunoblotting experiments with various antigens (summarised in Table 4).

The results of the immunoblotting (after SDS-PAGE) are visually presented

Table 1

Recovery of migrating A.suum and A.lumbricoides larvae from the liver of mice immunised with ES antigen of L2 larvae of either species three days after oral challenge with 2,000 embryonated eggs

Group (n = 5)	Immunised with	Challenged with	Number of larvae recovered individual values					mean per mouse	P-value (Wilcoxon)
I	A.suum	A.suum	0	5	5	9	12	6.2	vs IV < 0.05 vs VI < 0.05
II	A.suum	A.lumbricoides	0	3	4	6	8	4.2	vs V < 0.05 vs VII < 0.05
III	A.lumbricoides	A.lumbricoides	0	2	3	4	6	3.0	vs V < 0.05 vs VII < 0.05
IV	saline/adjuvant	A.suum	26	27	29	37	84	40.6	
V	saline/adjuvant	A.lumbricoides	19	37	41	53	67	43.4	
VI	non-treated	A.suum	36	39	42	51	58	45.2	
VII	non-treated	A.lumbricoides	24	28	33	42	54	36.2	

vs = versus

Table 2

Recovery of migrating A.suum and A.lumbricoides larvae from the lungs of mice immunised with ES antigen of L2 larvae of either species seven days after oral challenge with 2,000 embryonated eggs

Group (n = 5)	Immunised with	Challenged with	Number of larvae recovered individual values					mean per mouse	P-value (Wilcoxon)
I	A.suum	A.suum	4	6	0	0	8	3.6	vs IV n.s. VI < 0.05
II	A.suum	A.lumbricoides	0	3	0	2	0	1.0	vs V n.s. VII n.s.
III	A.lumbricoides	A.lumbricoides	0	0	3	0	0	0.6	vs V n.s. VII n.s.
IV	saline/adjuvant	A.suum	28	32	19	0	37	23.2	
V	saline/adjuvant	A.lumbricoides	0	15	10	16	8	9.8	
VI	non-treated	A.suum	30	18	25	14	23	22.0	
VII	non-treated	A.lumbricoides	20	8	19	0	12	11.8	

vs = versus
n.s. = not significant

154

Figure 1

SDS-Page of ES antigens of L2 larvae of Ascaris suum and immunoblotting
with immunesera of mice treated with ES antigens of L2 and L3/4 larvae
respectively

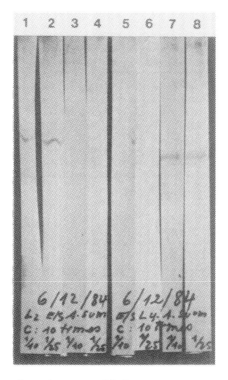

Antigens and antisera applied:

lane 1: ES/L2 A.suum, 10 times concentrated: immuneserum day 45 against
         ES/L2, 1 : 10 dilution
lane 2: ES/L2 A.suum, 10 times concentrated: against ES/L2, 1 : 25 dilution
lane 3: ES/L2 A.suum, 10 times concentrated: immuneserum day 45, against
         ES/L3/4, 1 : 10 dilution
lane 4: ES/L2 A.suum, 10 times concentrated: against ES/L3/4, 1 : 25
         dilution
lane 5: ES/L3/L4 A.suum, 10 times concentrated: immuneserum against ES/L2,
         day 45, 1 : 10 dilution
lane 6: ES/L3/L4 A.suum, 10 times concentrated, against ES/L2, 1 : 25
         dilution
lane 7: ES/L3/4 A.suum, 10 times concentrated, immuneserum against ES/L3/4,
         day 45, 1 : 10 dilution
lane 8: ES/L3/4 A.suum, 10 times concentrated, against ES/L3/4, 1 : 25
         dilution

Figure 2

SDS-Page of ES antigen of L2 larvae of Ascaris lumbricoides followed by immunoblotting using various immunesera from mice

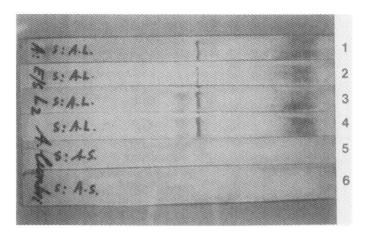

Antigens applied:
ES/L2 A.lumbricoides, concentrated 20 times in all lanes

Immunesera applied against:
lane 1: ES/L2 A.lumbricoides, day 45, dilution 1 : 10
lane 2: ES/L2 A.lumbricoides, day 45, dilution 1 : 25
lane 3: ES/L2 A.lumbricoides, day 40, dilution 1 : 10
lane 4: ES/L2 A.lumbricoides, day 49, dilution 1 : 25
lane 5: ES/L2 A.suum, day 49, dilution 1 : 10
lane 6: ES/L3/4 A.suum, day 49, dilution 1 : 10

Table 3

Results of serological examinations (by ELISA) on sera from mice immunised with ES/L2 antigens of A.suum and A.lumbricoides respectively and control animals after oral challenge infection with embryonated eggs of A.suum and A.lumbricoides at days 45 and 49

Ascaris antigens used in ELISA	Reciprocal ELISA antibody titers in pooled sera (n = 5) from mice immunised with						
	ES/L2 A.s *		ES/L2 A.s **		ES/L2 A.l **		Control
	day 45	day 49	day 45	day 49	day 45	day 49	days 45 and 49
ES/L2 A.s	80	80	80	80	160	160	< 20
ES/L2 A.l	20	< 20	20	20	160	160	< 20
ES/L3/4 A.s	20	< 20	20	20	< 20	< 20	< 20
Somatic L2 A.s	40	40	40	40	80	40	< 20
somatic L3/4 A.s	40	40	40	40	20	20	< 20

*   challenged with A.suum
** challenged with A.lumbricoides
A.s = Ascaris suum
A.l = Ascaris lumbricoides

Table 4

Recognition of ES antigens of Ascaris suum and Ascaris lumbricoides by Western Blotting after cross immunisation experiments.

Mouse antisera against	Ascaris antigens	ES/L2 A.s.	ES/L3,4 A.s.	ES/L2 A.l.
ES/L2	A.s.	2 bands	-	-
ES/L3,4	A.s.	-	1 band	-
ES/L2	A.l.	1 band	1 band	1 band

in the Figures 2 and 3. The sera of the mice immunised with ES/L2 antigens of A.suum produce a specific couple of two distinct lines representing homologous antigens with molecular weights of 47,4 and 47,2 K Dalton respectively. These lines were not detected with sera from mice immunised with ES/L3/4 antigens (animal experiment not presented here). Controversely, the latter reacted specifically with a homologous single antigen with a molecular weight of 46,7 KD (Figure 2), which was not recognised by the anti ES/L2 mouse sera.

As yet, no definite conclusions can be drawn from these figures. The ES antigens recognised are very close in molecular weights and are likely the main (protein) antigens which induce the protection as shown after challenge infection with either A.suum or A.lumbricoides.

Concentrated serum of mice immunised with ES/L2 A.lumbricoides antigen specifically reacted with an ES/L2 A.lumbricoides antigen with a molecular weight of 46,1 KD, which was not recognised by the sera of mice immunised with ES/L2 A.suum or ES/L3/4 A.suum antigens (Figure 3). These results are summarised in Table 4.

5. REFERENCES

1. Adams Oaks J and Kayes ST. Artificial Hatching and Culture of Toxocara canis. Second stage larvae. J.Parsitology 65, (6), 969-970, 1979.
2. Bindseil E. Immunity to Ascaris suum. I. Immunity induced in mice by means of material from adult worms. Acta Path.Microbiol. Scand. 77, 218-222, 1969.
3. Bindseil E. Immunity to Ascaris suum. II. Investigations of the fate of larvae in immune and non-immune mice. Acta Path.Microbiol. Scand. 77, 223-234, 1969.
4. Clegg JA and Smith MA. Prospects for the development of dead vaccines against helminths. Adv.Parasitol. 16, 165-218, 1978.
5. Crandall CA and Arean VM. The protective effect of viable and non-viable Ascaris suum larvae and egg preparations in mice. Am.J.Trop.Med.Hyg. 14, 765-769, 1965.
6. Glass WF, Briggs RC and Hnilica LS. Identification of tissue-specific nuclear antigens transferred to nitrocellulose from polyacrylamide gels. Science 211, 70-72, 1981.
7. Knapen F van Leusden J van, Polderman AM van and Franchimont JH. Visceral larva migrans: examinations by means of enzyme linked immunosorbent assay of human sera for antibodies to excretory-secretory antigens of the second stage larvae of Toxocara canis. Z.Parasitenkd 69, 113-118, 1983.
8. Leammli UK. Cleavage of structural proteins during the assembly of the head of bacteriophage T4. Nature (London) 227, 680-685, 1970.
9. Lloyd S. Progress immunisation against parasitic helminths. Parasitology 83, 225-242, 1981.
10. Ruitenberg EJ and Brosi BJM. Automation in Enzyme Immunoassay. Scand.J.Immunol. 8, suppl. 7, 63-72, 1978.
11. Savigny DH de. In vitro maintenance of Toxocara canis larvae and a simple method for the production of Toxocara ES antigen for use in serodiagnostic tests for visceral larva migrans. J.Parasitol. 61, (4) 781-782, 1975.
12. Stromberg BE and Soulsby EJL. Ascaris suum: immunization with soluble antigens in the guinea pig. Int.J.Par. 7, 287-291, 1977.
13. Stromberg BE, Khoury PB and Soulsby EJL. Development of larvae of Ascaris suum from the third to the fourth stage in a chemically defined medium. Int.J.Par. 7, 149-151, 1977.

14. Towbin H, Staehelin T and Grondon J. Electrophoretic transfer of
    proteins from polyacrylamide gels to nitrocellulose sheets: Procedure
    and some applications. Proc.Nat.Acad.Sci USA 76, 3116-3120, 1979.

TOXOCARA VITULORUM : A POSSIBLE AGENT OF LARVA MIGRANS IN HUMANS ?

K.VAN GORP, M.MANGELSCHOTS & J.BRANDT

1. INTRODUCTION

In recent years, serological surveys indicated repeatedly important sero-prevalences of Toxocara canis visceral larva migrans (VLM) in humans. As such, in the Federal German Republic (15) 6.4%, 0.5 to 5.7% in Japan (18) 7% in Dutch children (28) 2.6% in U.K. (21) 4.6 to 7,3% in U.S.A. (12), 5,1% in Switzerland (25) were reported, but most of these surveys indicate higher prevalences in certain population-groups. Because of the similarities in the life cycle of the bovine ascarid Toxocara vitulorum and T.canis, the possibility that the former could cause the VLM syndrome in humans seems to be realistic.

T.vitulorum is widespread in calves in Asia and Africa with a rather sporadic distribution in America and Europe, causing a high morbidity and mortality in buffalo calves (24). Faecal shedding of T.vitulorum eggs has been reported in calves as young as 6 days, but generally occurs at 19 days after birth, ceasing when calves reach approximately the age of 3 months (10, 11). Development towards mature worms starts mainly through transmammary infections (30).

The present study compares the infection in experimental hosts with previous reports in view of the occurence of T.vitulorum as VLM.

2. MATERIALS AND METHODS

Adult T.vitulorum were obtained after spontaneous expulsion from calves in Zambia. Females were dissected and the eggs stored in water with sulphuric acid 200 mmol/l at 28°C for 4 to 6 weeks.

Female CF1 mice, aged between 6 and 8 weeks received 2500 infective larvae per os. Mice were sacrificed at time intervals indicated in table 1, each time 3 infected and 1 control-mouse. The viscera were digested individually in 10mg/ml pepsine (1:10,000 Sigma) in water acidified with 11.4 mol/l hydrochloric acid until pH 1.7.

The same procedure was followed for Ascaris lumbricoides eggs, obtained after mebendazole treatment.

Simultaneously, second stage larvae of T.vitulorum and T.canis were maintained in vitro. Eggs were hatched by a combination of described methods (8, 9, 27). First, they were decoated with 6% v/v sodium hypochlorite for 45 min. and washed. The hatching medium was 2.5ml sodium bisulphite (0.1 mol/l) for 10 seconds then 2,5ml PBS (pH 7.2) and sodium bicarbonate (114 mol/l) plus 10 μl/ml hydrochloric acid (11.4 mol/l) was added.This mixture was left on a magnetic stirrer for 2 hours, washed and baermanised in RPMI-1640 (Flow Laboratories) with sodium bicarbonate (29 mol/l) plus 100 i.u./ml penicillin and 100 μg/ml streptomycin.

The same medium was used to maintain larvae, at a concentration of
10,000 larvae per ml, kept at 37°C with 5% carbon dioxide in air. In this
medium the larvae survived for at least two months. The supernatantia,
which are referred to as ES-Antigens were replaced once a week and those
taken during the second week in vitro, used in ELISA.

These metabolic antigens and a somatic extract of adult T.vitulorum were
tested against 4 sera from calves , shedding T.vitulorum eggs, sera from
calves known to be free of T.vitulorum or experimentally infected with
cysticercosis (serum dilution: 1/250).

3. RESULTS

The autopsy results of the mice infected with T.vitulorum second-stage
larvae are given in table 1. No larvae were recovered after day 103 PI.
The mean lenght of the larvae was 416 μm in the first week, 460 to 525 μm
in week 2 PI, it remained then steady at 600 μm, the last few larvae
(day 103 PI) measured 900 μm.

From table 2. the much shorter survival and larval recovery of A.lumbri-
coides in mice can be seen. No larvae were found in the digestiva after
day 16 PI, the range of their length was 950 μm (first week) - 1200 μm
(end of the second week).

The ELISA results are given in table 3.

4. DISCUSSION

Although the survival of VLM in experimental hosts is apparently depending on the hosting species e.g. Irfan (13) failed to find T.vitulorum
larvae in other viscera than the liver in guinea pigs, even the strain and
age seems to be important for T.canis VLM in mice (14). Nevertheless the
occurence of T.vitulorum larvae in the brain reflects some of the similarities with T.canis.

T.vitulorum larvae were present in the brain from day 4 until day 12 PI,
with the same appearance as described for T.canis: i.e. the absence or low
degree of cellular reactions (7) allowing for free movement of the larvae,
in fact, macroscopic haemorrhages, as described for T.canis in mice (3)
were equally observed in the present study.

An important difference with the reports on T.canis is the period during which larvae were observed in the brain : for T.canis larvae tend to
remain much longer in the brain, where their number is steadily increasing,
even when their appearance in other organs decreases but again the influence of the strain of mice involved seems to exist (7).

In the observations by Chauhan and Pande (6) with T.vitulorum in mice,
larvae were not encountered in the brain, whereas Warner (30) found only
one larva in the brain and this at 52 days PI. This observation and the
brain-dwelling larvae at day 96 PI (table 1) are probably exceptional
occurences.

Survival of T.canis in mice has been described by Olson (19) as being
well over 6 months. Numbers of T.canis larvae in the brain were still increasing on day 122 PI, but then already decreasing in other organs (7),
which led these authors to the contention that T.canis larvae in bitches
use the brain as a reservoir to escape the host-immune system.

Day PI	Intestines	Liver	Lungs	Carcass	Brain	Kidneys	Recovery	% Eosinophils
1	106.6(23.0)	538(267.0)	1.3(2.3)	40.7(28.4)	0	0	27.4%	0.5(0.5)
2	64(49.9)	272(195.7)	16(9.0)	10(3.4)	0	0	14.4%	2.7(1.6)
4	8(13.8)	152(48.9)	77(24.1)	18(0)	7(2.6)	7.6(9.3)	10.8%	4.7(2.1)
5	8(13.8)	64(36.6)	94(3.4)	56(13.8)	1.7(1.5)	20(6.9)	9.7%	9.3(4.5)
7	8(13.8)	20(13.8)	93.3(23.1)	140(36.1)	11(11.8)	12(12.0)	11.4%	9.5(3.3)
8	24(41.5)	28(28.3)	99.6(71.7)	132(57.2)	7.3(4.2)	15.3(7.6)	12.2%	15.5(14.3)
12	0	36(31.7)	82(21.0)	100(45.8)	0.3(0.6)	19.3(23.8)	9.5%+	8.7(1.7)
14	4(6.9)	20(13.8)	46(9.1)	150.6(37.8)	0	4(3.4)	9.0%	10.8(5.9)
19	0	6(6.0)	14(15.1)	38.6(46.1)	0	0	2.3%	13.5(1.3)
26	0	2.6(4.6)	0	21.3(23.1)	0	0	0.9%	6.6(0.1)
33	0	2.6(4.6)	4(6.9)	10(3.4)	0	0	0.6%	6.5(3.5)
96	4(6.9)	0	3.3(2.8)	53.3(5.0)	6(6.0)	0	2.7%	9(6.1)
103	0	0	0	4(3.4)	0	0	0.2%	8.5(0.7)
Mean No. of larvae per organ	226.6	1141.2	530.5	774.5	33.3	78.2		

+ +1 larva in the heart

Table 1: Mean number of T. vitulorum larvae (+ standard deviation) recovered per mouse after artificial digestion of the viscera. (Infection dose: 2500 larvae per mouse)

Day P.I.	Intestines	Liver	Lung	Carcass	Recovery	% Eosinophils
1	0	354 (52.7)	0	0	14.2%	1.2 (1.1)
2	0	457 (35.3)	0	0	18.3%	0
6	0	159 (190.9)	9 (4.2)	8 (11.3)	7.0%	4.2 (0.3)
10	54 (59.4)	12 (17.0)	15 (12.7)	60 (84.8)	5.6%	14.2 (11.0)
16	36 (25.4)	24 (34.9)	0	14 (2.8)	3.0%	2.0 (0.7)
32	0	0	0	0	0	2.0 (0.6)

Table 2. Mean number (+ standard deviation) of Ascaris lumbricoides larvae recovered per mouse after artificial digestion of the viscera.

(Infection dose : 2500 larvae per mouse. No larvae were recovered from the brain, kidneys, heart).

	Cows (with T.vit. positive off-spring)	Calves infected with T.vitulorum	Calves T. vitulorum-free
T.vitulorum Somatic Ag	1.18 (0.13)	0.47 (0.06)	0.28 (0.15)
T.vitulorum ES Ag	0.66 (0.16)	0.11 (0.03)	0.20 (0.10)
T.canis ES Ag	1.52 (0.01)	0.26 (0.06)	0.44 (0.24)

Table 3. Mean (s.d.) values of Toxocara canis and Toxocara vitulorum antigens testes by ELISA against sera from 4 calves with fecal shedding of T.vitulorum eggs, their dams and control sera (3 negative and 2 positive for cysticercosis).

The comparison between the results in table 1 with those by Prokopič and Figallová (20) suggests that T.vitulorum VLM behave rather like Toxocara cati VLM i.e. for the latter: survival in the brain until day 17, but starting on the first day PI, whereas for T.canis they observed brain-dwelling larvae from day 4 until day 18.

No larvae of A.lumbricoides were observed in the brain (table 2), similar with previous observations on A.lumbricoides (20) and Toxascaris leonina (17, 20).

The distribution of T.vitulorum larvae in the viscera and the carcass (table 1) seems to coincide with those reported for T.canis (19, 23). Levels of eosinophilia (table 1) remained fairly constant throughout the infection but not significantly higher than for A.lumbricoides (table 2), contrasting with the observations of Lukeš (16) where eosinophilia due to A.lumbricoides was always higher than when due to toxocariasis canis. If T.vitulorum VLM behave like T.canis or at least like T.cati, then the risk for man does exist, whereby the manipulation of bovine faeces as fertilizer, fuel or even as building material may add an additional factor to its epidemiology.

Whereas, in general, for T.canis the prenatal infection seems to be dominating over the transmammary route (4, 5), for T.vitulorum it seems to be the reverse (1, 30). The presence of the larvae in the milk of the cow, from 2 to 22 days post partum (26) has been indicated as a possible health hazard for humans (2). However, this might not be necessarily true since larvae infecting calves by this route, develop to maturity without leaving the digestive tract (30).

Although the high specificity of T.canis - ES antigen human serology has been described (22, 29) from the preliminary ELISA results (table 3) it seems obvious that further purification of T.vitulorum ES will be necessary to allow for distinction amongst the Toxocara species in man.

REFERENCES

1. Anwar AH and Chaudry AH: An insight into the life cycle of cattle ascarid. Pakistan Veterinary Journal, 4: 71-72, 1984.
2. Banerjee DP, Barman Roy AK and Sanyal PK: Public health significance of Neoascaris vitulorum larvae in buffalo milk samples. Journal of Parasitology, 69: 1124, 1983.
3. Bisseru B: Studies on the liver, lung, brain and blood of experimental animals infected with Toxocara canis. Journal of Helminthology, 43: 267-272, 1969.
4. Burke TM and Roberson EL: Prenatal and lactational transmission of Toxocara canis and Ancylostoma caninum: experimental infection of the bitch before pregnancy. International Journal for Parasitology, 15: 71-75, 1985.
5. Burke TM and Roberson EL: Prenatal and lactational transmission of Toxocara canis and Ancylostoma caninum : experimental infection of the bitch at midpregnancy and at parturition. International Journal for Parasitology, 15: 485-490, 1985.
6. Chauhan PPS and Pande BP: Migratory behaviour and histopathology of Neoascaris vitulorum larvae in albino mice. Indian Journal of Experimental Biology, 10: 193-200, 1972.

7. Dunsmore JD, Thompson RCA and Bates IA: The accumulation of Toxocara canis larvae in the brains of mice. International Journal for Parasitology, 13: 517-521, 1983.
8. Fairbairn D: The in vitro hatching of Ascaris lumbricoides eggs. Canadian Journal of Zoology, 39: 153-162, 1961.
9. Gupta AK: A simple method of artificial hatching of Toxocara canis in vitro and preparation of excretory-secretory antigen. Current Science, 53: 529-530, 1984.
10. Gupta SC: Pattern and control of Neoascaris vitulorum in calves. Indian Veterinary Journal, 63: 71, 1986.
11. Gupta SC and Suresh Singh K: Migratory behaviour of Neoascaris vitulorum (Goeze, 1782) in white mice. Indian Journal of Animal Science, 54: 517-518, 1984.
12. Herrmann N, Glickman LT, Schantz PM, Weston MG and Domanski LM: Seroprevalence of zoonotic toxocariasis in the United States. American Journal of Epidemiology, 122: 890-896, 1985.
13. Irfan M: Studies on the development of Neoascaris vitulorum. Pakistan Veterinary Journal, 4: 70, 1984.
14. Koizumi T. and Hayakawa J: Mouse strain difference in visceral larva migrans of Toxocara canis. Experimental Animals, 33: 291-295, 1984.
15. Lamina J, Hertkorn U and Künast C: Infektionen mit dem Hundespulwurm Toxocara canis als Erreger einer Larva migrans visceralis im Kindesalter. Pädiatrische praxis, 27: 293-301, 1982/83.
16. Lukeš S: Changes in the white blood picture during experimental larval ascariasis, toxocariasis and toxascariasis. Folia Parasitologica (Praha) 32: 237-245, 1985.
17. Matoff K and Komandarev S: Comparative studies on the migration of the larvae of Toxascaris leonina and Toxascaris transfuga. Zeitschrift für Parasitenkunde, 25: 538-555, 1965.
18. Matsumura K and Endo R: Seroepidemiological study on toxocaral infection in man by enzyme-linked immunosorbent assay. Journal of Hygiene (Camb.), 90: 61-65, 1983.
19. Olson LJ: Organ distribution of Toxocara canis larvae in normal mice and in mice previously infected with Toxocara, Ascaris or Trichinella. Texas Report on Biology and Medicine, 20: 651-657, 1962.
20. Prokopič J and Figallová V: Migration of some roundworm species in experimentally infected white mice. Folia Parasitologica (Praha), 29: 309-312, 1982.
21. Slovak AJM: Toxocaral antibodies in personnel occupationally concerned with dogs. British Journal of Industrial Medicine, 41: 419, 1984.
22. Speiser F and Gottstein B: A collaborative study on larval excretory/ secretory antigens of Toxocara canis for the immunodiagnosis of human toxocariasis with ELISA. Acta Tropica, 41: 361-372, 1984.
23. Sprent JFA: On the migration behaviour of the larvae of various Ascaris species in white mice. I. Distribution of larvae in tissues. Journal of Infectious Diseases, 90: 165-176, 1952.
24. Srivastava AK and Sharma DN: Studies on the occurence, clinical features and pathomorphological aspects of ascariasis in buffalo calves. Veterinary Research Journal, 4: 160-161, 1981.
25. Stürchler D, Bruppacher R und Speiser F: Epidemiologische Aspekte der Toxokariasis in der Schweiz. Schweizerische Medizinische Wochenschrift, 116: 1088-1093, 1986.

26. Tongson MS: Neoscaris vitulorum larvae in milk of Murrah buffalo. Philippine Journal of Veterinary Medicine, 10: 60-63, 1971.

27. Urban Jr JF, Douvres FW and Tromba FG: A rapid method for hatching Ascaris suum eggs in vitro. Proceedings of the Helminthological Society, 48: 241-243, 1981.

28. van Knapen F, van Leusden J en Conijn - van Spaendonk MAE: Toxocara infekties, diagnostiek en voorkomen bij de mens in Nederland. Tijdschrift voor Diergeneeskunde, 108: 469-474, 1983.

29. Van Knapen F, van Leusden J, Polderman AM and Franchimont JH: Visceral larva migrans: examinations by means of enzyme linked immunosorbent assay (ELISA) of human sera for antibodies to excretory-secretory (ES) antigens of the second stage larvae of Toxocara canis. Zeitschrift für Parasitenkunde, 69: 113-118, 1983.

30. Warren EG: Observations on the migration and development of Toxocara vitulorum in natural and experimental hosts. International Journal for Parasitology, 1: 85-99, 1971.

ANTIGENIC AND BIOCHEMICAL ANALYSES OF THE EXCRETORY-SECRETORY
MOLECULES OF *TOXOCARA CANIS* INFECTIVE LARVAE

B.D. ROBERTSON, S. RATHAUR, R.M. MAIZELS

1. ABSTRACT
Current knowledge of the origin, biochemistry and antigenic nature of
Toxocara canis ES is reviewed. Information is presented on a number of
functions, including acetylcholinesterase and protease activities, and
improved methods of diagnosis are discussed.

2. INTRODUCTION
The Excretory-Secretory or ES molecules released by parasitic nematodes
in vitro have been analysed for a number of species including *Ascaris suum*
(12), *Trichinella spiralis* (24), *Nippostrongylus brasiliensis* (3), *Nemato-
spiriodes dubius* (3), *Litomosoides carinii* (9), *Brugia pahangi* (17, 25),
Brugia malayi (11) and *Toxocara canis* (16, 37). Most, but not all (9, 25)
ES molecules have been shown to be antigenic *in vivo*, and those produced
by *Trichinella* (8) and *Toxocara* (22) provide some degree of protection in
immunised animals. ES forms the basis of serodiagnosis for toxocariasis
(5, 6, 32) and as such it is important to understand both its origins and
its interaction with the host. In this paper our current knowledge of the
nature and composition of *T. canis* ES will be reviewed and its possible
functions and role in the host immune response will be discussed.

3. ORIGIN OF ES
T. canis second-stage infective larvae are remarkable both for their
longevity *in vitro* (greater than one year in serum-free culture medium)
and their production of large amounts of ES antigens. Estimates of up to
200 pg ES per larvae per day have been made (16) which must involve a
considerable proportion of the available metabolic energy, in turn suggesting that the production of these molecules must be a high priority for
the parasite. However, the physiological origin of these molecules is not
clear. Radio-labelling studies have shown that all bar one component of
ES can be found on the larval surface (16).
However, it is not known if they all originate there, or are first
secreted from internal glands and then deposited to some extent on the
surface. Certainly if worms radiolabelled on the surface are subsequently
cultured, labelled molecules are released and appear in the culture medium.
Smith and coworkers (33) showed that specific antibody bound only transient-
ly to the surface of larvae unless worms were treated with azide on ice
to limit their ability to turn over surface molecules. It will be important
to use monospecific probes for these surface and secreted antigens to
investigate their site of synthesis and routes of transport in the whole
organism.

4. BIOCHEMICAL NATURE OF ES

The number of ES components reported in the literature varies between three (5) and fourteen (35). This variation probably results from several factors including parasite culture, ES collection protocols, the polyacrylamide gel electrophoresis (PAGE) systems and detection assays used combined with any heterogeneity found within parasite populations. In our work, we

kD	Radiolabelling	Protease sensitivity				Gel Staining	Lectin overlay
		pronase	pepsin	trypsin	V8P		
400	Iodo, B-H	+	–	–	+	PAS	HPA
120	Iodo,B-H,Met	+	–	–	–	PAS,silver	HPA
70	Iodo,B-H,Met	+	+	+	+	silver	Con A
55	Iodo,B-H,Met	+	+	+	–	silver	Con A
32	Iodo,B-H,Met	+	+	+	–	PAS,silver, CBB	Con A

FIGURE 1. Autoradiograph of Iodogen labelled ES separated by SDS-PAGE, including summary of biochemical data relating to individual bands. Abbreviations : Iodo, Iodogen mediated iodination of tyrosine residues; B-H, Bolton-Hunter mediated iodination of lysine residues; Met, ^{35}S-methionine metabolic labelling; PAS, Periodic Acid Schiffs; CBB, Coomassie brilliant blue; HPA, *Helix pomatia* agglutinin; Con A, Concanavilin A; V8P, *Staphylococcus aureus* V8 protease.

have focussed on the five major bands of 32, 55, 70, 120 and 400 K, and generally radiolabel ES proteins to achieve the maximum level of sensitivity. A typical profile obtained by SDS-PAGE of Iodogen labelled ES is shown in Figure 1.

All components bar the 400 K band label with ^{35}S-methionine (20, 38), indicating that this molecule has a very limited protein component, possibly of unusual composition. The biochemical nature of individual components has been examined by a combination of gel staining, lectin binding and enzyme degradation techniques (20), also summarised in Figure 1. Staining data indicates a high degree of glycosylation on many molecules and in general, the higher the molecular weight the greater the proportion of sugar to protein observed. Lectin overlay data show the smaller components (32, 55, 70 K) contain mannose, while the larger molecules (120 and 400 K) are bound by HPA specific for N-acetyl galactosamine, the predominant sugar found in carbohydrate analysis of the whole ES (20) as summarised in Table 1.

One interesting aspect of glycosylation, is the presence of human A and B blood group-like substances on both the larval surface and in the ES (34). Although their presence may be incidental to the host-parasite relationship, these determinants will be recognised by antibodies of similar specificity and may interfere with immunodiagnosis.

TABLE 1. CARBOHYDRATE ANALYSIS OF WHOLE ES

CARBOHYDRATE	% TOTAL CARBOHYDRATE
N-acetylgalactosamine	58.4
Galactose	24.5
Mannose	6.4
N-acetylglucosamine	4.8
Glucose	3.3
Fucose	2.4
Xylose	0.2

5. ANTIGENIC STRUCTURE OF ES

Apanel of eight monoclonal antibodies (Tcn1-8) made to ES have been used to characterise its antigenic structure (18), employing radio- immuno-precipitation and immunoblotting experiments to assess the occurence and nature of epitopes on both surface and ES antigens. Six of the monoclonals react with heat-stable carbohydrate determinants while the remainder react with periodate-resistant determinants and probably recognise peptide determinants. All the monoclonals bind to more than one component of ES, indicating a high degree of epitopic repetition which is further borne out when they are used in a two-site ImmunoRadioMetric Assay (IRMA) to assess which molecules bear epitopes recognised by more than one antibody (18). In addition, by using the same monoclonal antibody as capture and second antibody, the presence of repetitive epitopes could be discerned (Table 2).

TABLE 2. AMOUNT OF ^{125}I-ANTIBODY BOUND IN A TWO-SITE IRMA

		Capture antibody on plate							
		Tcn1	Tcn2	Tcn3	Tcn4	Tcn5	Tcn6	Tcn7	Tcn8
	Tcn1	-	-	-	-	-	-	-	-
	Tcn2	+	+++	++	+	+	++++	+	+
Radio-	Tcn3	+	+	+	+	+	+	+	+
labelled	Tcn4	++	++	+	+++	++++	+	++	+
second	Tcn5	++	++	+	++	++	+	++	+
antibody	Tcn6	-	-	-	-	-	-	-	-
	Tcn7	+++	++	-	++++	++++	-	++	+
	Tcn8	-	+	-	+	+	-	+	+

Table 2 shows that four of the antibodies (Tcn 2, 4, 5 & 7) bind repeated epitopes. The results obtained using different capture and second antibodies clearly show that some molecules bear several epitopes bound by different monoclonals for example, the molecule captured by Tcn 2 bears epitopes also recognised by Tcn 4, 5 and 7. The similar nature of the patterns obtained here and in immunoprecipitation experiments suggests that Tcn 4, 5 and 7 bind closely related determinants. Competitive inhibition experiments confirm this (18). Immunoprecipitation of surface labelled molecules by the monoclonals shows that they bear epitopes shared with ES molecules (18). However immunofluorescence on live parasites, suitably treated to minimize surface turnover, indicates that only the Tcn 2 and 8 determinants are accessible to extrinsic antibody at the parasite surface.

6. BIOLOGICAL FUNCTION OF ES MOLECULES

So far only the physicochemical nature of these antigens has been considered. Attention is now turning to the more fundamental question, both in terms of pathology and parasite survival, of the biological function of ES molecules.

Acetylcholinesterases (AChE) are activily secreted by a number of parasitic helminths including *N. brasiliensis* (13), *Trichostrongylus colubriformis* (23), *Setaria cervi, L. carinii, B. pahangi* (39)and *Schistosoma mansoni* (40). Recently AChE has been found in the ES of *T. canis* (28). The enzyme has now been partially purified using copper chelating (1) and Concanavilin A affinity chromatography. It is antigenic and cross reacts by ELISA with antibodies raised to electric eel (*Electrophorus electricus*) AChE. The major component of the partially purified fraction is the 70 K glycoprotein, although this may be copurifying under the same conditions as the enzyme. Substrate specificity is that of a true AChE and activity is inhibited by eserine (28).

Protease activity has been studied in the ES of several helminths including *S. mansoni* (14, 15),*Fasciola hepatica* (2), *B. malayi* (26) and *Ancylostoma caninum* (10) and has now been detected in the ES of *T. canis* (30). Preliminary studies using SDS-PAGE gels copolymerised with a gelatin substrate, as described by McKerrow *et al* (14), have localised proteolytic activity to 400 and 50-55 K. Activity is inhibited both in this and Azocoll assays by phenylmethyl-sulphonyl fluoride, consistent with serine protease activity. An *in vitro* assay involving degradation of an extracellular matrix (14) by live parasites suggests that most of this activity is directed against elastin. Such an enzymatic function for ES is not surprising, considering the extensive tissue migrations undertaken by the larvae in both paratenic (27) and definitive hosts (41), and further characterisation and purification of this enzyme is underway.

An interesting possibility is that some secreted products are able to modulate the host immune response in some way beneficial to the parasite. It has been postulated that parasite derived AChE may hydrolyse host acetylcholine, which is thought to play a role in regulating the immune system (29). Another example is eosinophilia, a major clinical manifestation of many helminth infections including toxocariasis (4). ES administered intraperitoneally to mice is able to induce eosinophilia in the absence of the parasite (36). The existence of a parasite-derived eosinophil-inducing factor is a novel observation which warrants further study and may provide information on the pathogenesis of not only *Toxocara* but also other parasitic helminths. Interestingly, although *Toxocara* elicit a strong eosinophilic response, eosinophils do not appear to be able to kill larvae. Thus, using eosinophils and serum from ocularly infected guinea pigs (31), or from a human toxocariasis patient (7), granulocytes bound to the surface of larvae and displayed all the cytoplasmic changes associated with activation, including degranulation. However the larvae showed little signs of damage and appeared to terminate the interaction by shedding an extracuticular layer to which the eosinophils had bound. This sloughing off of the outermost membrane may be à macroscopic version of the surface antigen release referred to above (16, 33).

7. IMPROVED IMMUNODIAGNOSIS

Toxocaral serodiagnosis is currently based on the detection of antibodies to ES, but as with many pathogens which elicit persistent humoral responses, this is not necessarily indicative of current infection. Two potential routes to improve immunodiagnosis are open. One is to assay for circulating ES antigen in infected host blood. Matsumura *et al* (19) described a sandwich ELISA using a polyclonal rabbit antiserum against ES which successfully detected toxocaral antigen in the serum of infected dogs, finding highest antigen levels in the serum of naturally infected puppies one month old. Although there was little cross-reactivity and low intra-assay variation, the sensitivity was not very high (minimum 100ng ES). In this laboratory a two-site IRMA (42) using monoclonal antibody Tcn 2 as both capture and visualising layers, has proved of some use in detecting free ES antigen in the serum of experimentally infected mice and rabbits (30). Antigen detection would also seem to be restricted by the relatively short time span over which antigen is free in the circulation, prior to its immune-mediated clearance or complexing to specific antibody. Thus, in preliminary experiments, circulating antigen in mice could be detected, using Tcn 2 - 2, up to nine days post infection; after this time antibody levels rose dramatically and even after acid-treatment to dissociate immune complexes, no further antigen could be detected (30).

The second method involves the detection of antibodies to specific epitopes. This could be achieved in either of two ways. Firstly a highly sensitive competitive assay can be used in which monoclonal antibody binding to ES or parasite extract is inhibited by test antibody, such as that described by Mitchell *et al* (21) for *Schistosoma japonicum*. One of the attractions of this approach is that a complex antigen need not be fractionated before coating plates. Microtitre plates are coated with appropriate antigen (in this case ES), and incubated with a mixture of test serum and radio- or enzyme- linked monoclonal antibody. Inhibition of monoclonal antibody binding gives an index of the amount of host antibody to that epitope present in the serum. Ideally the epitope chosen should be one which does not possess blood group activity, does not cross-react with other parasitic helminths, and which is only seen in current, as opposed to past, infections. Alternatively, a cloned ES peptide with these characteristics could be used, bringing with it the added advantage of an invariant, precisely defined antigen in unlimited quantities. Both these options are currently being investigated.

ACKNOWLEDGEMENTS.

This work was supported by the Medical Research Council, the WHO/TDR and The Wellcome Trust.

172

REFERENCES

[1]   Bjerrum OJ, Selmer J, Hangaard J & Larsen F: Isolation of human erythrocyte
        acetylcholinesterase using phase separation with Triton X-114 and monoclonal
        immunosorbent chromatography. Journal of Applied Biochemistry 7:356-369, 1985
[2]   Chapman CB & Mitchell GF: Proteolytic cleavage of immunoglobulin by enzymes released
        by *Fasciola hepatica*. Veterinary Parasitology 11:165-178,1982
[3]   Day KP, Howard RJ, Prowse SJ, Chapman CB & Mitchell GF: Studies on chronic versus
        transient intestinal infections in mice. 1. A comparison of responses to
        excretory/secretory (ES) products of *Nippostrongylus brasiliensis* and *Nematospiroides
        dubius* worms. Parasite Immunology 1:217-2391979
[4]   Dent JH, Nicholas RL, Beaver PC, Carrera GM & Staggers RJ :Visceral Larva Migrans
        with a case report. American Journal of Pathology 32:777-803,1956
[5]   De Savigny DH: In vitro maintenance *Toxocara canis* larvae and a simple method for the
        production of *Toxocara* ES antigen for use in serodiagnostic tests for visceral larva
        migrans. Journal of Parasitology 61:781-782, 1975
[6]   De Savigny DH, Voller A & Woodruff AW: Toxocariasis: serological diagnosis by enzyme
        immunoassay. Journal of Clinical Pathology 32:284-288,1979
[7]   Fattah DI, Maizels RM, McLaren DJ & Spry CJF:*Toxocara canis* : Interaction of human
        eosinophils with the infective larvae. Experimental Parasitology 61: 421-431, 1986
[8]   Gamble HR: *Trichinella spiralis*: Immunization of mice using monoclonal antibody
        affinity-isolated antigens. Experimental Parasitology 59:398-404,1985
[9]   Harnett W, Meghji M, Worms MJ & Parkhouse RME: Quantitative and qualitative changes
        in production of excretions/secretions by *Litomosoides carinii* during development in
        the jird *(Meriones unguiculatus)*. Parasitology 93: 317-331,1986
[10]   Hotez PJ, Le Trang N, McKerrow JH & Cerami A: Isolation and characterization of a
        proteolytic enzyme from the adult hookworm *Ancylostoma caninum*. Journal of
        Biological Chemistry 26:7343-7348,1985
[11]   Kaushal NA, Hussain R, Nash TE & Ottesen EA: Identification and characterization of
        excretory-secretory products of *Brugia malayi* adult filarial parasites. Journal of
        Immunology 129:338-343,1982
[12]   Kennedy MW & Qureshi F: Stage-specific secreted antigens of the parasitic larval stages
        of the nematode *Ascaris*. Immunology 58:515-522,1986
[13]   Lee DL: The fine structure of the excretory system in adult *Nippostrongylus brasiliensis*
        (Nematoda) and a suggested function for the excretory glands. Tissue & Cell
        2:225-231,1970
[14]   McKerrow JH, Keene WE, Jeong KH & Werb Z: Degradation of an extracellular matrix by
        larvae of *Schistosoma mansoni*. 1. Degradation by cercariae as a model for initial
        parasite invasion of the host. Laboratory Investigation 49: 195-200,1983
[15]   McKerrow JH, Pino-Heiss S, Lindquist R & Werb Z: Purification and characterization of
        an elastinolytic proteinase secreted by cercariae of *Schistosoma mansoni*. Journal of
        Biological Chemistry 260:3703-3707,1985
[16]   Maizels RM, De Savigny D & Ogilvie BM: Characterisation of surface and
        excretory-secretory antigens of *Toxocara canis* infective larvae. Parasite Immunology
        6:23-37,1984
[17]   Maizels RM, Denham DA & Sutanto I: Secreted and circulating antigens of the filarial
        parasite *Brugia pahangi*: Analysis of *in vitro* released components and detection of
        parasite derived productd *in vivo*. Molecular and Biochemical Parasitology 17:277-288,
        1985
[18]   Maizels RM, Meghji M, Kennedy MW, Robertson BD & Smith HW: Expression of shared
        and specific epitopes on the surface and secreted antigens of the parasitic nematode
        Toxocara canis. submitted for publication,1987

[19]   Matsumura K, Kazuta Y, Endo R & Tanaka K: Detection of circulating toxocaral antigen in dogs by sandwich enzyme-immunoassay. Immunology 51:609-613, 1984

[20]   Meghji M & Maizels RM: Biochemical characterisation of larval excretory-secretory (ES) glycoproteins of the parasitic nematode *Toxocara canis*. Molecular and Biochemical Parasitology 18:155-170,1985

[21]   Mitchell GF, Premier RR, Garcia EG, Hurrell JGR, Chandler HM, Cruise KM, Tapales FP & Tiu WU: Hybridoma antibody-based competitive ELISA in *Schistosoma japonicum* infection. American Journal of Tropical Medicine and Hygiene 32:114-117,1983

[22]   Nicholas WL, Stewart AC & Mitchell, GF: Antibody responses to *Toxocara canis* using sera from parasite-infected mice, and protection from toxocariasis by immunization with ES antigens. Australian Journal of Experimental Biology and Medical Science 62:619-626,1984

[23]   Ogilvie BM, Rothwell TLW, Bremner KC, Schnitzerling HJ, Nolans J & Keith RK: Acetylcholinesterase secretion by parasitic nematodes. I Evidence for secretion of the enzyme by a number of species. International Journal of Parasitology 3:589-597,1973

[24]   Parkhouse RME & Clark NWT: Stage specific secreted and somatic antigens of *Trichinella spiralis*. Molecular and Biochemical Parasitology 9:319-327, 1983

[25]   Parkhouse RME, Clark NWT, Maizels RM & Denham DA: *Brugia pahangi*: labelling of secreted antigens with ^{35}S-methionine *in vitro*. Parasite Immunology 7:665-668,1985

[26]   Petralanda I, Yarzabal L & Peissens WF: Studies on filarial antigens with collagenase activity. Molecular and Biochemical Parasitology 19:51-59,1986

[27]   Prokopc J & Figallova V: Migration of some roundworm species in experimentally infected white mice. Folia Parasitologia (Praha) 29:309-313,1982

[28]   Rathaur S, Robertson BD, Selkirk ME & Maizels RM. unpublished data

[29]   Rhoads ML: Secretory cholinesterases of nematodes: Possible functions in the host-parasite relationship. Tropical Veterinarian 2:3-10, 1984

[30]   Robertson BD, McKerrow JH, Bianco AE, Selkirk ME & Maizels RM unpublished data

[31]   Rockey JR, Donnelly JJ, McKenzie DF, Stromberg BE & Soulsby EJL: *In vitro* interaction of eosinophils from Ascarid-infected eyes with *Ascaris suum* and *Toxocara canis* larvae. Investigations in Opthalmology and Visual Science 24:1346-1357,1983

[32]   Schantz PM, Meyer D & Glickman LT: Clinical, serologic and epidemiologic characteristics of occular toxocariasis. American Journal of Tropical Medicine and Hygiene 28:24-28,1979

[33]   Smith HV, Quinn R, Kusel JR & Girdwood, RWA: The effect of temperature and anti-metabolites on antibody binding to the outer surface of second stage *Toxocara canis* larvae. Molecular and Biochemical Parasitology 4:183-193, 1981

[34]   Smith HV, Kusel JR & Girdwood RWA: The production of human A and B blood group like substances by *in vitro* maintained second stage *Toxocara canis* larvae: their presence on the outer larval surfaces and in their excretions/secretions. Clinical and Experimental Immunology 54:625-633, 1983

[35]   Speiser F & Gottstein B: A collaborative study on larval excretory-secretory antigens of *Toxocara canis* for the immunodiagnosis of human toxocariasis with ELISA. Acta Tropica 41:361,1984

[36]   Sugane K & Oshima T: Induction of peripheral blood eosinophilia in mice by excretory and secretory antigen of *Toxocara canis* larvae. Journal of Helminthology 58:143-147,1984

[37]   Sugane K & Oshima T: Purification and characterization of ES antigen of *Toxocara canis* larvae. Immunology 50:113-120,1983

174

[38]   Sugane K, Howell MJ & Nicholas WL: Biosynthetic labelling of the excretory and secretory antigens of *Toxocara canis* larvae. Journal of Helminthology **59**: 147-151, 1985

[39]   Swamy KHS & Subrahmanyan D: Studies on some biochemical aspects of filarial parasites. Tropical Medicine and Parasitology **37**:92,1986

[40]   Tarrab-Hazdai R, Levi-Schaffer F, Smolansky M & Arnon R: Acetylcholinesterase of *Schistosoma mansoni* : antigenic cross- reactivity with *Electrophorus electricus* and its functional implications. European Journal of Immunology **14**:205-209,1984

[41]   Webster GA: On prenatal infection and the migration of *Toxocara canis* Werner, 1782 in dogs. Canadian Journal of Zoology **36**:435-440,1958

[42]   Zavala F, Cochrane AH, Nardin EH, Nussenzweig RS & Nussenzweig V: Circumsporozoite proteins of malaria parasites contain a single immunodominant region with two or more identical epitopes. Journal of Experimental Medicine **157**:1947-1957,1983

OCULAR TOXOCARIASIS: ROLE OF IgE IN THE PATHOGENESIS OF THE SYNDROME AND DIAGNOSTIC IMPLICATIONS.

C. GENCHI, P. FALAGIANI, G. RIVA, C. SIOLI.

ABSTRACT

Toxocariasis is a zoonotic parasitic infection of man caused by the larval stage of Toxocara canis, a common parasite of dogs. The clinical disorders caused by the migration of nematode larvae through the human tissues are currently grouped into two main syndromes: visceral larva migrans (VLM) and ocular larva migrans (OLM). The ocular lesions seem to be caused by a small number of larvae that may escape the immunosurveillance mechanisms of the host. This hypothesis is confirmed by the inconstant rises in IgG values observed by ELISA in sera of patients with suspected OLM. In this paper we report the detection of specific IgE and IgG in sera and aqueous fluid from selected OLM patients. This study was designed to compare the sensitivity of the two classes of antibodies, IgE detected by RAST and IgG by an ELISA method. In both cases we used an excretory-secretory T. canis antigen obtained from in vitro cultures of second stage larvae. The results show that IgE antibodies are always present in both sera and in the aqueous fluid while IgG are often low and were not measurable in one case in serum or aqueous fluid. Moreover, the higher levels of IgE in the aqueous fluid of same OLM patients compared with IgE levels in other patients with inflammatory syndromes (aspecific uveitis) may demonstrate the in situ production of IgE against the parasite and emphasize the specific role of these antibodies in the pathogenesis (IgE-mediated inflammatory reactions) and in the diagnosis of OLM syndrome.

INTRODUCTION

Toxocariasis is a zoonotic parasitic infection of man with the larval stage of Toxocara canis, a common parasite of dogs. The clinical disorders caused by migration of nematode larvae through human tissues are courrently grouped into two main syndromes, visceral larva migrans (VLM) and ocular larva migrans (OLM). Intraocular involvement causes important lesions and although the precise incidence of the disease is not established, the diagnosis is being made with increasing frequency (16).

It has been suggested that ocular toxocariasis is caused by a small number of larvae that have escaped the immunosurveillance mechanisms of the host (7, 15). This hypothesis is supported by the absence of eosinophilia and inconstant rises in IgG values observed in patients with suspected OLM

176

(15, 12, 9).

In preliminary observations, we demonstrated IgE antibodies in sera
of both VLM and OLM patients (1, 7). We concluded that this class of immunoglobulins is very specif and sensitive for diagnosis of OLM syndrome,
and less so for the VLM (6). However it is possible that the small amounts
of antigen circulating in patients with the ocular syndrome can stimulate
the production of larva-specific IgE rather than IgG.

In this paper we report the detection of larva-specific IgE and IgG
in sera and aqueous fluid from selected OLM patients. The study was designed to compare the sensitivity and specificity of this class of antibod
ies and to define the role of IgE in the pathogenesis of human toxocariasis, especially the ocular syndrome.

MATERIALS AND METHODS

Sera. Sera from 17 patients with suspected OLM and 12 with VLM were
collected and alquots were frozen at -20°C until used. From 9 OLM patients
we also obtained aqueous fluid. OLM cases with unilateral eye disease
otherwise undefined etiology were selected on the basis of funduscopic
picture (18, 3). VLM cases were selected in accordance with the criteria
of Glickman et al. (10) and De Savigny et al. (4); one was also checked
hystologically (Table 3, patient V10).

Antigen. The second stage T. canis larvae exoantigen (TEX) was used.
The eggs were obtained by dissection of uteri of adult female worms and
incubated in 1% aqueous formalin for 30 days at 27°C. The embrionated eggs
were hatched as described in the method of Matsumura et al. (13). The
hatched larvae were cultured sterirelly in MEM (Minimal Essential Medium,
Flow Laboratories) containing 1% glutamine, for 30 days, with the medium
renewed every 10 days, as described by De Savigny (4). The culture medium
containing excretory and secretory antigens was collected, centrifuged, fil
tered (Millipore 0.45 µm) and finally dialyzed against deionized water.
The final solution, containing 70 ug/ml of protein (Lowry method), was
stored at -35°C until used.

The antigens were analyzed by crossed immunoelectrophoresis (CIE)
(11) with the specific antiserum against TEX-antigen raised in the rabbit.

ELISA-IgG. The method used was the indirect microplate enzyme-linked
immunosorbent assay described by Voller et al. (19), slightly modified (2).
Polystyrene 96-well microtiter plates (Greiner) were coated by passive absorption overnight at 4°C with 100 µl of TEX-antigen diluted to 0.7 µl/ml
in carbonate buffer (pH 9.6) in each well. The plates were washed four
times with phosphate-buffered saline (pH 7.4) containing 0.05% Tween 20
(PBS-Tween).

Sera were diluted in PBS-Tween and 100 ul volumes were added to each
well. After an overnight incubation at 4°C, the plates were washed to remove unbound serum components and the wells were then filled with 100 µl
of horseradish peroxidase-conjugated goat antiserum against human IgG
(Miles-Yeda Ltd.) diluted 1:1,500 in PBS-Tween, and incubated for 1 hour

at 37°C. After a final washing, 50 µl of orthophenylenediamina in phossphate-citrate buffer (pH 5.5) containing H_2O_2 (0.015%) were added to each well. The reaction was stopped after 30 minutes by addition of 50 µl of 4N H_2SO_4. The results were read photometrically at 490_{nm} with a microplate reader and expressed as optical density (o.d.$_{490nm}$). The blank for every plate was determined by reading the optical density of a well not containing serum, but otherwise fully processed. The sera were routinely diluted 1:1,000.

RAST-IgE. The test was performed by the technique first described by Wide et al. (20), with some modification. Briefly, polystyrene beads coated with TEX-antigen (Allergosfere[R] Lofarma, Italy) were incubated on a horizontal shaker for $3\frac{1}{2}$ hours with 50 µl of serum. The beads were washed 3 times with saline caontaining 0.4% Tween 20. 50 µl of ^{125}I-anti-IgE (Fc specific) (Sferikit[R], Lofarma, Italy) solution were added to each tube, and beads incubated overnight on a horizontal shaker. The beads were then washed 3 times, as above, and the radioactivity measured in a gamma counter. The bound radioactivity percent of the total added radioactivity was calculated.

Total IgE. Total IgE were determined by PRIST[R] method using the manufacturer's instruction (Pharmacia Diagnostics, Uppsala, Sweden).

Depletion of IgG. This was carried out by affinity chromatography with protein A from Staphylococcus aureus Cowan strain, which binds subclasses 1, 2 and 4 of human IgG. Briefly, a chromatography column Ø 1 cm x 3.5 cm was filled with Protein A-Sepharose CL-4B (Pharmacia AB, Sweden), following the manufacturer's directions. The column was equilibrated with 0.15 M PBS, pH 7.2, containing 0.06% NaN_3. A sample of 0.35 ml serum was applied to the column and run through the gel for two hours. Finally, the serum was eluted from the column, concentrated to the original volume (0.35 ml) and the column washed with several volumes of PBS. The adsorbed IgG was then eluted from the column with 1 M acetate-buffer, pH 2.4. The protein peak (o.d.$_{280nm}$) was collected and the pH immediately neutralized by adding the proper quantity of concentrated (10 x) PBS, then dialyzed against PBS and concentrated to the original volume (0.35 ml).

Three sera were depleted of IgG, two from VLM patients, the third from a patient suspected of VLM, who gave a positive reaction for specific IgG (o.d. 0.80$_{490nm}$) but a negative one for specific IgE (0% bound radioactivity).

RESULTS

The results are summarized in tables 1-4. Specific IgE were detectable in both sera and aqueous fluid from OLM patients, while specific IgG were sometimes low and were not measurable in serum or in aqueous fluid of two cases (Table 1, samples 01 and 015).

The comparison of total IgE levels in sera and in aqueous fluids from OLM patients and those from subjects with aspecific uveitis showed significantly more antibody in aqueous fluid from OLM patients, the lev-

178

els of serum IgE were almost the same in both group (table 2).

TABLE 1. Specific IgG and IgE in sera and aqueous fluids from OLM patients.

| Patients | Sera | | Aqueous | |
	IgG TEX-ELISA o.d.490nm	IgE TEX-RAST % bound radioactivity	IgG TEX-ELISA o.d.490nm	IgE TEX-RAST % bound radioactivity
0 1	0.03	2.4 (+)	n.t.	23.2 (4+)
0 2	0.50	6.3 (2+)	n.t.	n.t.
0 3	1.00	15.0 (3+)	1.20	15.2 (3+)
0 4	1.13	24.8 (4+)	n.t.	12.4 (2+)
0 5	0.60	11.2 (3+)	1.40	25.5 (4+)
0 6	0.75	17.5 (3+)	1.00	18.3 (3+)
0 7	0.45	6.0 (2+)	n.t.	n.t.
0 8	0.60	18.5 (2+)	n.t.	n.t.
0 9	1.53	19.1 (3+)	n.t.	n.t.
010	0.50	5.2 (2+)	n.t.	n.t.
011	0.45	8.3 (2+)	n.t.	n.t.
012	1.34	5.7 (2+)	1.38	6.3 (2+)
013	0.60	15.8 (3+)	n.t.	n.t.
014	0.55	7.2 (2+)	n.t.	n.t.
015	0.45	11.2 (3+)	0.08	29.3 (4+)
016	1.50	36.0 (4+)	1.60	38.0 (4+)
017	1.30	2.7 (1+)	0.85	23.1 (3+)

TEX-ELISA cut off 0.45
n.t.: not tested

TABLE 2. Total IgE in sera and aqueous fluids from patients with suspected OLM or aspecific uveitis.

| | OLM | | | Aspecific uveitis | | |
patients	serum	aqueous		patients	serum*	aqueous*
0 3	260 IU	400 IU		AU 1	170 IU	5 IU
0 5	180 IU	600 IU		AU 2	270 IU	5 IU
0 6	280 IU	3700 IU		AU 3	95 IU	5 IU
012	200 IU	1200 IU		AU 4	750 IU	15 IU
016	450 IU	720 IU		AU 5	180 IU	5 IU
				AU 6	180 IU	15 IU
				AU 7	120 IU	5 IU

* both sera and aqueous fluids were negative to specific TEX-ELISA and
TEX-RAST.

In the sera from VLM patients, high titres of specific IgG were observed (> 1.90 o.d.$_{490nm}$) while IgE were low (Table 3). After depletion of IgG, IgE antibodies were significantly higher (Table 4). In serum number X1 (Table 4), from a patient suspected of VLM, positive for specific IgG but not for specific IgE, tests for this class of immunoglobulins were also negative after IgG depletion.

TABLE 3. Specific IgG and IgE in sera from VLM patients.

patients	IgG TEX-ELISA o.d.$_{490nm}$	IgE TEX-RAST % bound radioactivity
V 1	> 2.50	9.8 (2+)
V 2	1.90	4.8 (1+)
V 3	> 2.50	2.5 (1+)
V 4	> 2.50	6.5 (1+)
V 5	1.90	13.1 (2+)
V 6	> 2.50	32.3 (4+)
V 7	> 2.50	17.1 (2+)
V 8	> 2.50	6.2 (1+)
V 9	> 2.50	30.6 (4+)
V10	2.00	4.4 (1+)
V11	> 2.50	30.6 (4+)
V12	> 2.50	9.8 (2+)

TEX-ELISA cut off 0.45

TABLE 4. Specific IgE in sera from VLM* before and after depletion of IgG

Sera	Before IgG depletion TEX-ELISA o.d.$_{490nm}$	TEX-RAST % bound radioactivity	After IgG depletion TEX-ELISA o.d.$_{490nm}$	TEX-RAST % bound radioactivity	Eluted from the column TEX-ELISA o.d.$_{490nm}$	TEX-RAST % bound radioactivity
V 6*	> 2.50	32.3 (3+)	0.0	50.7 (4+)	1.20	0.0
V 9*	> 2.50	4.4 (1+)	0.0	38.2 (4+)	1.50	0.0
X 1	0.80	0.0	0.0	0.0	0.60	0.0

Antigen analysis by CIE revealed the presence of three distinct antigens in TEX-antigen, two of which migrate towards the anode, one towards the cathode. The calcium medium used for the cultivation of the larvae, tested as a control with the same technique, and the antiserum did not contain any bands (Figure 1).

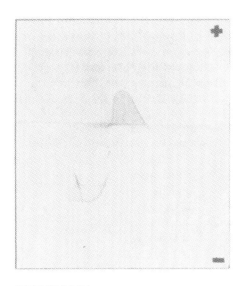

Fig. 1 CIE analysis of an extract of I. canis excreto-ry/secretory antigen (45 ug) with a rabbit antiserum at a concentration of 15 ul/cm^{-2}. Anode at top and right.

DISCUSSION

The present study confirms the specificity and sensitivity of the TEX- antigen for diagnosis of toxocarial syndromes in human beings.

Specific anti-Toxocara IgE can be detected in sera from both OLM and VLM patients. The notable increase in the amounts of these antibodies in VLM sera after IgG depletion suggests that large amounts of IgG might mask the presence of IgE in an in vitro system like ELISA. We hypotesize that there is a competition between the two classes of immunoglobulins for the same epitopes of the parasite. However, the interference is not enough to prevent detection of IgE in the serum.

The low levels of specific IgG sometimes found in our OLM patients, correlated with good specific IgE positivity, seems to confirm that small number of larvae responsible for the ocular syndrome probably cause pre-dominantly an IgE response. On the other hand, the higher total IgE levels observed in aqueous fluids from OLM patients than in those from aspecific uveitis indicate a local production (intraocularly) of this class of anti-bodies, as advanced by Rockey et al. (14). If our study be confirmed by further observations, we can conclude that IgE antibodies have a predomi-nant role in the pathogenesis of OLM syndromes. Binding of IgE antibodies to mast cells and basophils sensitizes these cells to release mediators that induce an inflammatory process in the choroid. This phenomenon can seriously damage the eye structures, as demonstrated in experimentally sensitized guinea pigs (14).

Our results do not provide a final answer to the many problems of immunolgy and pathogenesis of ocular toxocariasis, which has been shown by many investigators to be very complicated (21, 14, 15, 12, 5) it is our opinion that IgE is an important component in the response of the human host against the invasion of Toxocara larvae. In the eye, the amplifying

effects of IgE on the immune response may be dramatic and in addition dete
ction of this class of antibodies, along with IgG, can make the diagnosis
of ocular syndromes more reliable.

ACKNOWLEDGEMENTS
We thank G. Carlevaro, MD, of the Department of Ophtalmology,
S. Gerardo dei Tintori Hospital, Monza, for the clinical diagnoses of OLM;
M. Scaglia, MD, and M. Tinelli MD, of the Department of Infectious Dis-
eases, Policlinico S. Matteo, Pavia, for the clinical diagnoses of VLM;
dr. G. Mistrello for the technical help.
This work was supported by a grant M.P.I. 40%.

REFERENCES
1.   Brunello F., Genchi C., Falagiani P., 1983. Detection of larva-specif-
      ic IgE in human toxocariasis. Transaction of the Royal Society of
      Tropical Medicine and Hygine, 77, 279.
2.   Brunello F., Falagiani P., Genchi C., 1986. Enzyme immunoassay (ELISA)
      for detection of specific IgG antibodies to Toxocara canis ES anti-
      gens. Bollettino Istituto sieroterapico milanese, 65, 55-60.
3.   Carlevaro G., Spallone A., Ortolina S., Ridling P., 1986. Ocular
      toxocariasis diagnosed by RAST of the aqueous and the serum. IV Inter
      national Symposium on Immunopathology of the Eye. Padova, in press.
4.   De Savigny D.H., Voller A.,Woodruff A.W., 1979. Toxocariasis: serolo-
      gical diagnosis by enzyme immunoassay. Journal Clinical Pathology,
      32, 284-288.
5.   Ghafoor S.Y.A., Smith H.T., Lee W.R., 1984. Experimental ocular toxo-
      cariasis: a mouse model. British Journal of Ophthalmology, 68, 86-96.
6.   Genchi C., Brunello F., Falagiani P., Sioli C., Almaviva M., 1984.
      Larva specif IgE against Toxocara canis hatched larvae antigens in
      sera of toxocaral syndromes patients and healthy blood donors, in I.
      de Carneri (Ed) "Immunological diagnosis and other diagnostic methods
      for parasitic infections", report of a Conference sponsored by The
      First European Congress of Clinical Microbiology, Bologna 17th-21st
      October, 27-29.
7.   Genchi C., Brunello F., Boero M., Piantino P, Sioli C., Falagiani P.,
      1986. Antibodies against T. canis in patients with suspected syndrome
      of larva migrans. Annali dell'Istituto Superiore di Sanità, 22, 473-
      476.
8.   Glickman L.T. and Schantz P.M., 1981. Epidemiology and pathogenesis
      of zoonotic toxocariasis. Epidemiologic Reviews, 3, 230-250.
9.   Glickman L.T., Grieve R.B., Lauria S.S., Jones D.L., 1985. Serodiag-
      nosis of ocular toxocariasis: a comparison of two antigens. Journal
      Clinical Pathology, 38, 103-107.
10.   Glickman L.T., Schantz P.M., Dombroske R., Cypess R., 1978. Evaluation
      of serodiagnostic tests for visceral larva migrans. American Journal
      Tropical Medicine and Hygine, 27, 492-498.
11.   Grouba A.O., 1983. Crossed immunoelectrophoresis. Scandinavian Jour-

nal of Immunology, 17, suppl. 10, 113-124.

12. Kielar R.A., 1983. Toxocara canis endophthalmitis with low ELISA titer. Annals of Ophthalmology, 15, 447-449.

13. Matsumura K. and Endo R., 1981. Evaluation of hatching methods for Toxocara canis. Journal of Veterinary Medicine, 8, 61-62.

14. Rockey J.H., Donnelly J.J., Stromberg B.E., Laties A.M., Soulsby E.J. L., 1981. Immunopathology of ascarid infection of the eye. Archives of Ophthalmology, 99, 1831-1840.

15. Searl S.S., Moazed K., Albert D.M., Marcus L.C., 1981. Ocular toxocariasis presenting as leukocoria in a patient with low ELISA titer to Toxocara canis. Ophthalmology, 88, 1302-1306.

16. Schantz P.M., Weis P.E., Pollard Z.F., White M.C., 1980. Risk factors for toxocaral ocular larva migrans: a case control study. American Journal Public Health, 70, 1269-1271.

17. Shields J.A., 1984. Ocular toxocariasis. A review; Survey of Ophthalmology, 28, 361-381.

18. Vinciguerra P., Carlevaro G., Spallone A., 1986. Le granulome choroidien peripherique secondaire a une toxocarose oculaire. Bulletin mensuel de la Société Française de Ophtalmologie, 97, 67-70.

19. Voller A., Bidwell D., Bartlett A., 1976. Enzyme immunoassays in diag nostic medicine. Bulletin of the World Organisation, 53, 55.

20. Wide L., Bennich H., Johansson S.G.O., 1967. Diagnosis of allergy by an in vitro test for allergen antibodies. Lancet, 2, 1105.

21. Zinkham W.H., 1978. Visceral larva migrans, a review and reassessment indicating two forms of clinical expression: visceral and ocular. American Journal Diseases of Children., 132, 627-633.

SEROLOGICAL ARGUMENTS FOR MULTIPLE ETIOLOGY OF VISCERAL LARVA MIGRANS

J.C.PETITHORY, F.DEROUIN, M.ROUSSEAU, M.LUFFAU and M.QUEDOC.

ABSTRACT

The authors give five biological and epidemiological criteria for the diagnosis of Visceral Larva Migrans Syndrome (VLMS) as also the preliminary results of serodiagnostic tests using antigens of various nematodes, mainly Ascaris suum, Toxocara canis and Nippostrongylus brasiliensis.

Two hundred embryonated eggs of Toxocara canis were administered to "two mentally defective infants" by SMITH and BEAVER and they developed eosinophilia which persisted for more than 13 months (18). This experiment proved that T.canis larvae may be a cause of VLMS and fostered the idea that it was mostly if not only caused by T.canis (24-8).

A systematic study using numerous antigens of nematodes for serodiagnosis of VLMS allows us to give some supportive evidence in favour of the multiplicity of etiology of VLMS and of the partial role of T.canis.

1. MATERIAL AND METHODS

ANTIGENS

Origin: The adults of Ascaris suum, Parascaris equorum, Ascaridia galli and Toxocara vitulorum were freshly collected at a slaughterhouse. It is to be noted that pig parasitism by A.suum has strongly declined in recent years because of the systematic anthelminthic treatment. We could not find T.vitulorum any more in spite of repeated requests to numerous abattoirs in Paris region, Normandy and Brittany, some of which are specialized in calf slaughter.

Adults of Ascaris lumbrocoides were collected after treatment of human cases.

Toxocara canis : A strain of this ascarid has been maintained in our laboratory since 1967. The bitches are periodically fed with infective eggs of T.canis mixed in food. The adult worms are then collected from one month or one and a half month old pupies. Attempts to infect adult dogs did not succeed and this has a relation to the work of SPRENT (19).

The intestines of 87 adult dogs used for transplant studies were opened. T.canis was found in the intestines of 2 dogs; one with 3 adults, the other one with approximately 20. Our laboratory strain comes from this source. Toxascaris leonina was found in 19 and Trichuris vulpis in 11 dogs. The adult worm antigens were prepared from these worms.

The larval excretory-secretory (ES) antigen of T.canis was prepared according to a technique similar to that of DE SAVIGNY (7) using RPMI as culture media.

The adults of Nippostrongylus brasiliensis and Haemonchus contortus came from strains maintained at the I.N.A. (11).

The adults of <u>Toxocara cati</u> came from cats captured at random and those of <u>Porrocaecum ensicaudatum</u> from blackbirds from Oise area.

The larvae of <u>Anisakis simplex</u>,mainly L4 stage, were collected in herrings coming from the North Sea (15).

<u>Preparation</u> : Freshly collected adults worms and larvae are washed repeatedly in saline and frozen. They are then crushed 6 times with successive freezing and thawing. The extraction of hydrosoluble antigens is made in 1 % NaCl prepared in bidistilled water. The antigen solution is dialysed for 24 hrs at 4°C, then lyophilized. The antigen is then titrated against several sera of variable reactivity and eventually used in 2 different dilutions if zone phenomena should appear: as is the case with <u>A.suum</u>.

<u>METHODS</u>

The micro-Ouchterlony, electrosyneresis and immunoelectrophoresis are performed in agarose gel, Veronal-tris buffer pH 9,2, fi 0,05.

The micro-Ouchterlony wells are refilled with sera and antigens after 24 hrs at room temperature then kept during 4 days at 4°C.

The immunoelectrophoresis are generally performed with serum concentrated 3 times by lyophilisation.

The slides are washed with stirring for 4 hrs in a solution of sodium citrate at 5% to eliminate the reactions of protein C substance (10), they are then washed, dried and stained with amidoschwartz.

<u>PATIENTS</u>

Between 1966 and 1986 sera of four thousand patients were tested for visceral larva migrans syndrome (VLMS). They usually were studied against <u>A.suum</u> and <u>T.cati</u> adult worm antigens, except for a short period of time because of a shortage of <u>T.canis</u> antigen. Other antigens such as <u>P.equorum</u>, <u>T.vitulorum</u>, <u>T.cati</u>, <u>T.leonina</u> were also systematically used during shorter period of time.

DIAGNOSTIC CRITERIA FOR VLMS.

Most of the time a parasitological proof of VLMS cannot be provided. In an unusual host the causative larvae cannot reach adult stage and hence the nematode eggs are never found in stool. The recovery of the larvae from the patient is quite impossible because of their small size and numbers and their search in hepatic biopsy material may be futile. The larvae have usually been found in enucleated eyes. Except in some rare cases of heavy infections there is no mortality. Thus it is exceptional that the larvae could be found back in tissues in histological sections or after peptic digestion on autopsy.

The lack of parasitological proof in such cases implies that there should be strict diagnostic criteria, especially in experimental studies. The diagnosis, particularly the immunodiagnosis, cannot be based only on serology as regards to the choice of positive cases. In our present study we deliberately excluded 177 sera that were positive against <u>T.canis</u> and/or <u>A.suum</u> but for which we lack one of the following criteria such as no stool examination was performed or the patient did not stay in a tropical country. For the physician, these criteria may be less strict and lead to a probable diagnosis permitting treatment. The five following biological and epidemiological criteria are compulsory for a scientific study of their sera.

1) Eosinophilia over 500/mm3

A high blood eosinophilia is almost always found, except in 2 cases which we have consequently excluded. This eosinophilia is explained by tissue contact of the parasite and its lack of adaptation to man. In cases of ocular larva migrans, usually one or just a few larvae are localized in the eye. Because of an insufficient amount of antigen or more probably because of the isolated position of the eye and the absence of exchange with the other parts of the body, the larvae in eyes give no general reactions though eosinophils may be detected in the aqueous component of the eye.

2) Positive serology against one or several nematodes antigens

The serology is most often negative in the ocular larva migrans syndrome, however, the reactions are positive if the aqueous component of the eye is used.

3) Negative Fasciola hepatica serology or weaker than the nematode antigen serology

In 127 cases of F.hepatica infections which were serologically positive we obtained 41 (32%) positive reactions using A.suum antigen. Out of 10, 9 were positive also against T.canis antigen. In such cases, the reactions against nematode antigens are always less strong than against F.hepatica antigen. Because of the similarity in the clinical pictures, the late appearance and the difficulty in recovery of F.hepatica eggs, the diagnosis of VLMS may be drawn (or has been drawn) and serology to diagnose fascioliasis is not performed.

4) Absence of human nematodiasis

The stool examination should be negative for nematode eggs. This criterion can be discussed since the intestinal nematodiasis diagnosed by usual techniques give negative serological reactions. However if several nematodes such as Ancylostoma and Ascaris are associated, we have noticed faintly positive serological reactions. Anisakiasis gives cross reactions with A.suum and T.cati (15).

5) The patient has never lived in a tropical country

The nematodiasis, mainly filariasis very often found in tropical countries, show cross reactions with various nematode antigens. In this context onchocercosis is important as with T.canis adult worm antigen we have found in Ouchterlony 57 (67%) positive reactions out of 84 cases.

Strongyloïdiasis, with its cycle of auto-infection and its great frequency of occurrence in tropical countries, may also be responsible for cross reactions though less frequently than in case of filariasis. For these reasons the patients who have lived in tropical countries should be excluded from serological study involving diagnosis of VLMS. That does not mean this syndrome is absent in such areas. On the contrary, because of the conditions of climate and hygiene, it is probably more frequent than in temperate areas, but the serological diagnosis is often camouflaged by numerous other human nematode infections. From a public health point of view the importance of VLMS is negligible with regard to human nematodiasis in tropical countries.

2. RESULTS

Antigens of A.suum and T.canis

Out of 292 sera of VLMS, 262 (90%) were found positive against the adult worm antigen of A.suum in Ouchterlony gel diffusion test.

Out of the same 292 sera, 109 (37,5%) were found positive against the adult worm antigen of T.canis in Ouchterlony technique. The preliminary study with the ES antigen of T.canis by Ouchterlony (42 sera) gave positive results in approximately 43% of the tested sera. In about half of the cases the positive sera for T.canis ES antigen were negative for adult worm antigen of T.canis.

All things considered, the sera positive for adults and/or ES antigens of T.canis correspond to about 60% of the studied sera.

Can nematodes other than T.canis play a role in the VLMS? Presently we shall restrict ourselves to N.brasiliensis. A detailed study on the other nematodes antigens bringing complementary data in favour of the multiple etiology of the VLMS will be published elsewhere.

Antigen of N.brasiliensis

With the adult worm antigen of N.brasiliensis a survey has been carried out on 207 sera from patients with VLMS diagnosed by Ouchterlony method against antigens of A.suum and/or T.canis.

81 out of 207sera gave positive results; 73 gave one precipitin band, 7 gave two precipitin bands, 1 gave three precipitin bands and in 34 (14%) of these patients the number of precipitin bands was equal or even superior to those obtained with the other nematodes antigens. In 60 (74%) of these seropositive patients, we have in addition a positive response to H.contortus antigen.

NAME	Eosinophilia	N. brasiliensis Ad (Ou)	N. brasiliensis Ad (I)	H. contortus Ad	A. suum Ad (Ou)	A. suum Ad (I)	T. canis Ad	T. canis L	T. cati Ad	T. leonina Ad	P. equorum Ad	T. vitulorum Ad	A. galli Ad	A. simplex L	T. spiralis L	Loa loa L	F. hepatica Ad
RO	48%	1	1	0	0	0	0	0	1	1+	0	0	0	0	0	0	0
PE	56%	1	2	1	0	1+	0	0	0	0	0	0	0	0	0	0	0
LE	42%	1	1	0	0	0	0	0	0	0	0	0	0	0	0	0	0

TABLE I: Serological study of three patients with eosinophilia over 35% and negative for A.suum, T.canis and F.hepatica antigens

Ad = adult L = larva Ou = Ouchterlony I = immunoelectrophoresis

ILLNESS	Onchocerciasis	Tropical eosinophilia	Ascaridiasis	Trichinosis	Ancylostomiasis	Strongyloïdiasis (Creeping)	Strongyloïdiasis	Distomatosis	Tumor	Cirrhosis	Syphilis	Various
Total number	12	6	6	6	6	4	6	6	4	4	3	15
Positives	3	2	0	1	0	1	0	2	0	0	1	1[+]

TABLE II: Specificity of the crude antigen of adult
Nippostrongylus brasiliensis by Ouchterlony.

[+]Salmonella septicemia.

The study of the specificity of the adult antigen of N.brasiliensis is given in Table II.

70 patients showing an hypereosinophilia of more than 35%, who had never been out of France or Europe and had no allergy and had a negative serology against T.canis, A.suum, F.hepatica adult antigen and T.spiralis larval antigen, were systematically tested against N.brasiliensis antigen by Ouchterlony method.

Three patients respectively aged   28, 48, 50 years were found positive against N.brasiliensis antigen. A complementary study against the other nematodes antigens has shown (Table I) one precipitin band by Ouchterlony against T.cati (RO), one precipitin band against H.contortus for PE and no precipitin band against the other 12 nematode antigens for LE. Their syphilitic serology was negative.

3. DISCUSSION

TOXOCARA CANIS

The role played by T.canis in the VLMS is certain, but for some epidemiologic reasons its importance can be discussed. The eggs are not embryonated immediately after leaving the host and therefore these are not infectious by direct contact with the dogs. Dog faeces are never used as a fertilizer so that the food contamination is rare. The contamination by geophagy is generally observed in young children, mentally defective infants and in cases of pica.

The studies showing the frequency of this parasite are numerous but give variable results. T.canis has been found among 21% English dogs but the

author adds "other unidentified ova, in retrospect almost certainly
T.leonina, were seen in the stools of 6 of the 300 animals" (24). We found
it among 3% of adult French dogs and experimentally proved the difficulty
of infecting them. We have to particularly keep in mind the indisputable
results by SPRENT (19) in Australia; 29 pupies aged between 1 and 6
months were all found carriers, and out of the 29 adults dogs, 3 were also
carriers with respectively 4, 1 and 1 adults worms. It implies that just
one out of these three adult dogs could possibly eliminate fertile eggs.
The spread of T.canis eggs in the environment is usually possible only
through the intermediary of pupies and therefore the potential remain limited. The serological data are also in favour of a limited role played by
T.canis. The reactions against adult or larval antigens of T.canis are
positive in only 40% of the cases and in 60% if we include the two stages.
   The large number of common antigenic fractions shared between T.canis
and T.cati; 16 out of 18 (3), and the similarity in the mode of infection
make it difficult to define the respective roles of these two ascarids in
the causation of VLMS.

ASCARIS SUUM
   In 1973, we suggested (13) the possible involvement of A.suum in VLMS
because of the large number of positive results against A.suum antigens.
In the present study out of 292 sera 90% are found positive which is a
high percentage. However, its role in the VLMS seems unlikely for the
following reasons:
   -PETTER C. (16) considers that the monoxenous Ascaridia such as
P.equorum, T.vitulorum and A.suum are non-pathogenic for man and if they
are accidentally ingested they will not migrate in the tissues.
   -Certain studies show the possibility for A.suum reaching adult stage
in man (6, 12, 21) and this led to propose the name, A.lumbricoides var.
suum (21), for the swine ascarid.
   -In 4 cases of heavy human infection, the immunoelectrophoresis was
negative for the 2 and the other 2 showed positive reaction with one band
of precipitin (17). Two human volunteers were infected; one with 8 embryonated eggs of A.suum, the other one with 16. On the third week a slight
though temporary eosinophilia appeared. They were followed during 6 months
and the serology was always negative (1).
In cases of weak or even heavy infections the absence of positive serology
is in favour to show that the evolution of A.suum in man is similar to
that of A.lumbricoides and against its role in VLMS.
   The A.suum antigen gives positive cross reactions in numerous other human nematodiasis particularly in filariasis (13) and anisakiasis (15).

NIPPOSTRONGYLUS BRASILIENSIS
   It is a cosmopolitan parasite and exists in France (5). The rat may
contaminate human food with its excrements. Epidemiologically speaking,
contamination of human adult is possible or even easier by this nematode
than the contamination by T.canis through the ingestion of soil or sand.
The study of the 3 cases showing hypereosinophilia and giving positive
reactions against N.brasiliensis antigen, bring some arguments in favour
of the role played by this worm.
- the patient (RO) has shown a febrile bronchopneumonia compatible with
the pulmonary migration of N.brasiliensis larvae.

- The study of N.brasiliensis antigen specificity shows cross reactions against nematodiasis such as onchocercosis, tropical pulmonary eosinophilia and creeping in anguillulosis. These illnesses are excluded for the three patients. Besides, trichinosis is serologically and epidemiologically excluded as the three patients were studied before the recent French epidemic. The syphilitic serology was negative and they did not present any septicemia.
- The serum of the patient (LE) reacted with none of the 13 nematodes antigens except against N.brasiliensis antigen.
However, the existence of cross reactions against other nematode antigens can not be excluded, as they are so numerous. The repeated negative serological reactions against adult and larvae of T.canis allow us to exclude this etiology.
It is to be noted that in our serological study on VLMS 39% were positive against N.brasiliensis and 38% against T.canis adult worm antigens. Thus the role played by N.brasiliensis in the etiology of some cases of VLMS is not certain but possible.

OTHER NEMATODES
The infective embryonated eggs or larvae of numerous species of nematodes either coming from domestic or wild animals may be accidently ingested by man. They can be either destroyed by the gastric juice or lead to the VLMS (14).
The finding and identification of nematodes other than T.canis implicated in the VLMS is difficult though few studies have been done;
- large size larva (620 μm x 57 μm) (20),
- larva twice the size of the one of T.canis (2) and
- larvae of Baylisascaris procyonis, raccoon ascaride (9)

IDENTICAL REACTIONS
As far back as 1928, CANNING (4) using the complement fixation test experimentally proved the existence of antigenic relations of A.suum with P.equorum, T.canis, A.galli and A.columbae.
A considerable sharing of antigens between larval stage and the adult worm was observed for T.canis and A.suum. Antisera absorbed with extracts of adults of T.canis and T.leonina detected "genus specific antigens in the third stage larvae and in the adults of A.suum but not in eggs and second stage larvae" (23).
BIGUET et al.(3) using immunoelectrophoresis technic and a rabbit immune sera anti A.suum showed 20 precipitin bands against the homologous antigen, 19 against A.lumbricoides, 18 against P.equorum, 12 against T.canis, 8 against T.cati, 5 against A.galli and 6 against A.columbae.
The importance of these cross reactions, also studied in other numerous works, prevents us in conclusively defining the respective role of each nematode by serology. The importance of reactions appreciated in Ouchterlony and immunoelectrophoresis by the number of precipitin bands and the importance of reactions with the larval antigens is of course an important argument to define the etiological role of the parasite. But a sound proof is lacking in these instances.

4. CONCLUSION
The first results of this serological study show that the etiologic role of T.canis in VLMS is certain but limited. It is not sure if A.suum, in

spite of 90% positive serological reactions using a homologous antigen, plays an etiological role.

Some nematodes giving positive serological reactions, such as N.brasiliensis, can be ingested by man at the infective stage and cause VLMS. The importance of cross reactions in serology requires us to be careful in our conclusions.

REFERENCES

1. Aubree P: Larva migrans viscerale: rôle de Ascaris suum. Thèse med., Paris 1973.
2. Beaver PC, Bowman DD: Ascaridoid larva (nematode) from the eye of a child in Uganda. American Journal of Tropical Medicine and Hygiene, 33, 1272-1274, 1984.
3. Biguet J, Rose F, Capron A, Tran Van Ky P: Contribution de l'analyse immunoélectrophoretique à la connaissance des antigènes vermineux. Incidences pratiques sur leur standardisation, leur purification et le diagnostic des helminthiases par immunoélectrophorèse. Revue d'Immunologie (Paris), 29, 5-23, 1965.
4. Canning GA: Precipitin reactions with various tissues of Ascaris lumbricoides and related helminths. American Journal of Hygiene, 9, 207-226, 1929.
5. Chabaud AG et Desset MC: Nippostrongylus rauschi n.sp.nematode parasite de dermoptères et considérations sur N.brasiliensis parasite cosmopolite des rats domestiques. Annales de Parasitologie (Paris), 41, 243-249, 1966.
6. Crewe W, Smith DH: Human infection with pig Ascaris (Ascaris suum). Annals of Tropical Medicine and Parasitology, 65, 85, 1971.
7. De Savigny DH: In vitro maintenance of Toxocara canis larvae and a simple method for the production of Toxocara ES antigen for use in serodiagnostic tests for visceral larva migrans. Journal of Parasitology, 61, 781-782, 1975.
8. Ehrhard T, Kernbaum S: Toxocara canis et toxocarose humaine. Bulletin de l'Institut Pasteur, 77, 225-287, 1979.
9. Kazacos KR: Raccoon ascarids as a cause of larva migrans. Parasitology today, 2, 253-255, 1986.
10. Longbottom JL, Pepys J: Pulmonary aspergillosis: diagnostic and immunological significance of antigens and C substance in Aspergillus fumigatus. Journal of Pathology and Bacteriology, 88, 141-151, 1964.
11. Luffau G: Propriétés biologiques du sérum hyperimmun provenant de rats infestés par Nippostrongylus brasiliensis. Annales de Recherches Vétérinaires, 3, 319-346, 1972.
12. Lysek H: Epidemiologia y especificidad de la ascaridiosis en el hombre y el cerdo. Revista Cubana de Medicina Tropical, 26, 3-24, 1974.
13. Petithory J, Brumpt L: Diagnostic immunologique des filarioses. Nouvelle Presse médicale, 2, 2059, 1973.
14. Petithory J, Brumpt LC: Diagnostic d'une hyperéosinophilie parasitaire. Information médicale et paramédicale (Montreal), 28, 1-6, 1976.
15. Petithory JC, Lapierre J, Rousseau M, Clique MT: Diagnostic sérologique de l'anisakiase (granulome éosinophile digestif) par précipitation en milieu gélifié (Ouchterlony, électrosynérèse et immunoélectrophorèse). Médecine et Maladies Infectieuses, 3, 157-162, 1986.

16. Petter C: Etude zoologique de la larva migrans. Annales de Parasitologie, 35, 118-137, 1960.

17. Phills JA, Harrold AJ, Whiteman GV, Perelmutter L: Pulmonary infiltrates, asthma and eosinophilia due to Ascaris suum infestation in man. New England Journal of Medicine, 286, 965-970, 1972.

18. Smith MHD, Beaver PC: Persistance and distribution of toxocara larvae in tissues of children and mice. Pediatrics, 12, 491-497, 1953.

19. Sprent JFA: Observation on the development of Toxocara canis in the dog. Parasitology, 47, 184-209, 1957.

20. Summer D, Tinsley EGF: Encephalopathy due to visceral larva migrans. Journal of Neurology, Neurosurgery and Psychiatry, 30, 580-584, 1967.

21. Takata I: Experimental infection of man with ascaris of man and the pig. Kitasato Archives of Experimental Medicine, 23, 49-52, 1951.

22. Voelckel J, Le Gonidec G, Jacoby JC, Jehl R: Helminthes hépato digestifs parasites de rattus norvégiens à Marseille. Médecine Tropicale, 24, 531-536, 1964.

23. Williams JF, Soulsby EJL: Antigenic analysis of the developmental stages of Ascaris suum. I. Comparison of eggs, larvae and adults. Experimental Parasitology, 27, 150-162, 1970.

24. Woodruff AW, Thacker CK: Infection with animal helminths. British Medical Journal, 1, 1001-1005, 1964.

EXPERIMENTAL TRICHINELLA SPIRALIS INFECTION IN TWO HORSES

F.VAN KNAPEN[1], J.H.FRANCHIMONT[1], W.M.L.HENDRIKX[2] and M.EYSKER

ABSTRACT

Recently two large epidemics of trichinellosis in man due to the consumption of infected horse meat, were described in France. Immediate control measurements were taken by the veterinary authorities in France and other European countries to prevent future infections of man and to start epidemiological surveillance of horses in general to get informed about this infection. A reliable serodiagnostic method for horse trichinellosis had first to be established. For this purpose, two horses were experimentally infected with T.spiralis, with 230 larvae per kg each. Both animals actively ate, without hesitation, the infected mouse material with their food. No clinical signs were observed during the infection period, nor did temperature rise in the first four weeks after the oral infection. An ELISA system was worked out to follow the antibody profile. Both animals rapidly produced specific sero-antibodies to T.spiralis. 10 weeks respectively 20 weeks after the infection, the animals were slaughtered and blood was collected as reference serum for the future. The distribution of T.spiralis larvae in various muscle tissues was measured as well. The parasites were predominantly found in tongue, masseter musculature and diaphragm, whereas very few were found in heart material. The animal slaughtered at 20 weeks proved very heavily infected with an average number of more than 100 larvae per gram of meat.

1. INTRODUCTION

Outbreaks of human trichinellosis in Europe normally are caused by the consumption of infected pig meat or meat from wild boars (11, 12). In the past decade however, human infection due to infected horse meat was reported twice (3, 4). Recently two large "urban type" outbreaks occurred in France (1) where over 1500 individuals became infected due to consumption of imported horses from eastern Europe and the United States respectively. No meat inspection regulations exist for the control of T.spiralis infections in horses. Immediate adaptation of an artificial digestion method for pig meat (8, 13) was introduced at abattoirs in France to prevent human infection. The same measurements were requested from countries exporting horse meat into France. Little or no information exists as to (i) whether horses can become naturally infected with T.spiralis (ii) how often horses are infected with T.spiralis and (iii) if a certain degree of infection (numbers and distribution of parasites) may be of direct public health consequence. In this short communication we report the oral infection of two horses with T.spiralis infected mouse material and the clinical follow-up of the two infected animals. Furthermore, a specific enzyme linked immunosorbent assay was developed and the antibody profiles of the infected horses were studied. At last both animals were slaughtered and the parasite distribution was measured in the carcasses.

2. MATERIALS AND METHODS

2.1. Horses

One male horse of about four years old at a weight of 325 kg was bought at the free market. The animal was in healthy condition except for blindness in one eye. Moreover the animal had a one side cryptorchy. A pony (130 kg) was obtained from the Veterinary Faculty (Institute of Parasitology and Helminthology) which was held there for several years. The parasitological status of this animal was followed over a long period and the animal was virtually free from parasites. Both animals were transported to the Institute and housed separately in two clean stables.

2.2. Infection experiment

After a one day starvation both animals were fed with pellets on top of which half an T.spiralis infected mouse was laid. The mice were enviscerated and skinned and cut in the length. The other halves of the mice were digested totally to measure the number of T.spiralis larvae. Both horses ate all the food including the mouse material without any hesitation. They were infected in this way with 230 larvae per kg body weight.

2.3. Clinical observation

Rectal temperature was measured daily after the infection. Apetite, behaviour, diarrhea and appearance of oedema was checked and noted daily by the animal caretakers.

2.4. Enzyme linked immunosorbent assay (ELISA)

A straight forward ELISA was worked out (7) using a crude saline extract of T.spiralis muscle larvae as antigen, and a commercially available peroxidase conjugated IgG fraction Goat anti-horse IgG (obtained from Cappel, Malvern, PA, USA). Checker board titrations were performed to find the optimal dilutions of the conjugate. As substrate 5-amino-2- hydroxy benzoic acid was used as described earlier (10) for pig sera.

2.5. Serum samples

Pre-infection sera were taken, and post-infection blood samples were collected with weekly intervals. Since no reference sera were available the preinfection serum and serum obtained at slaughter from the horse (10 weeks after infection) were used as control sera. Sera were obtained from horses, ponies and foals with a variety of disorders, from the Veterinary Faculty (n = 125) (see Table 1).

Table 1

The presence of antibodies against T.spiralis in The Netherlands

	Number of animals	ELISA results negative	postive
Horse (Vet.Faculty)*	50	50	0
Pony (Vet.Faculty)	22	21	1
Foal (Vet.Faculty)	53	53	0
Slaughter horse (Poland)	32	32	0

* Animals suffered from a wide variety of disorders: pharyngitis, bronchitis, pneumonia, cornage, skin problems, obstipation and colic, arthritis and other low motion disorders, neurological disorders and hepatitis. 13 animals were scrupiously tested for parasitic infection. Parasites found: small and large Strongylids, Oxyuris, Anaplocephala, Parascaris, Habronema muscae, Gastrophilus intestinalis.

At an abattoir 32 blood samples were collected from slaughterhorses imported from Poland.

2.6. Slaughtering
  The horse was slaughtered 10 weeks after infection and the pony at 20 weeks. Blood was collected in large quantities to prepare reference serum. The carcasses were scrupiously inspected for inbedded Strongylus species. Both S.vulgaris and S.edentatus were collected for antigen preparation. Small intestine (+ 1 meter) was collected to search for adult T.spiralis in the laboratory. Pieces of musculature for artificial digestion were collected from: diaphragm, heart, tongue, oesophagus, M.rectus abdominus, M.intercostalis, M.biceps, M.quadriceps, M.masseter and M.sterno mandibularis.

3. RESULTS

3.1. Horses

After the oral infection no clinical signs were observed besides a short lasting diarrhea of the horse in the 6th week. No changes in rectal temperature (see Figure 1) nor loss of apetite was observed. Both animals were slaughtered in perfect healthy condition.

3.2. ELISA

After some checker board titrations a straight forward ELISA was ready for use (see Figure 2). The follow-up study of the horse and the pony is presented in Figures 3 and 4, respectively. Both animals rapidly developed antibody titers reaching their maximum at 5-6 weeks after the infection. Both animals still had elevated titers when sacrificed at 10 and 20 weeks post infection respectively. Horses (ponies, foals) with other disorders did not show antibodies to T.spiralis when tested in this assay (Table 1). One pony reacted strongly positive. This pony originated the Institute of Parasitology and Helminthology, thus parasitological information was available, however the animal was sacrificed months ago. The pony harboured various small strongylids and both S.vulgaris and S.edentatus. Furthermore, it uniquely hosted the small strongylid Cylicodontophorus imparidentatum. 32 Sera from polish horses all reacted negative in ELISA/T.spiralis (Table 1). Strongylus vulgaris and Strongylus edentatus were found in large numbers in the horse and few of them in the pony at slaughter. Antigens were prepared (crude saline extracts) to be tested in ELISA with the sera obtained throughout the experimental period. No differences were observed in ELISA during the course of the T.spiralis infection indicating that no cross-reaction between the two species (Trichinella and Strongylus) exists.

Approximately one meter of small intestine of the horse (10 weeks after infection) was cut into small portions and examined in a Baermann method to search for adult T.spiralis. No adults were recovered. Histopathology of the intestine showed no indication of inflammation or presence of adult parasites. The intestine of the pony (20 weeks after infection), therefore, was not tested for adult T.spiralis. The results of the artificial digestion of different muscles are presented in Tables 2 and 3. A remarkable difference in degree of infection between the two animals was observed. However, in both the distribution was comparable, i.e. most parasites were found in M.masseter, tongue and diaphragm, whereas the smallest numbers were found in the heart.

Table 2

Results tissue digestion T.spiralis pony at 20 weeks post-infection

Muscle tissue	Muscle bundle in weight/grams	Total number of larvae	Number of larvae/gram
Diaphragm	140	10210	72,93
Heart	190	122	0,64
Tongue	100	6000	60,00
M.rectus abdominus	150	5090	33,93
Oesophagus	26	218	8,38
M.intercostalis	126	1183	9,39
M.biceps	290	5580	19,24
M.quadriceps	300	2740	9,13
M.masseter	200	24380	121,90
M.sterno mandibularis	76	603	7,93

Table 3

Results tissue digestion T.spiralis horse at 10 weeks post-infection

Muscle tissue	Muscle bundle in weight/grams	Total number of larvae	Number of larvae/gram
Diaphragm	400	229	0,57
Heart	300	14	0,05
Tongue	100	80	0,80
M.rectus abdominus	100	10	0,10
Oesophagus	50	10	0,20
M.intercostalis	60	30	0,50
M.biceps	400	38	0,10
M.quadriceps	400	27	0,07
M.masseter	200	190	0,95
M.sterno mandibularis	200	22	0,11

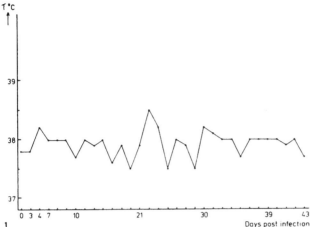

Follow-up of rectal temperature (C°) in an experimentally with
T. spiralis infected pony

FIGURE 1

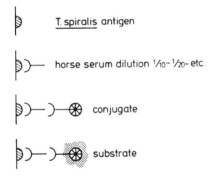

ELISA for T. spiralis antibodies in horse serum.

T. spiralis antigen

horse serum dilution ⅟₁₀- ⅟₂₀- etc.

conjugate

substrate

FIGURE 2

- Crude T.spiralis antigen is diluted in sodium carbonate buffer (0.1 M,
  pH 9.6) at a dilution of 5 µg/ml total protein and incubated for 1 hour
  at 37°C.
- After washing, two fold horse serum dilutions, starting 1 : 10, prepared
  in 0.01 M PBS, pH 7.2 with sheep serum (diluted 1 : 150) and 0.05 %
  Tween 20 are added.
- Incubation occurrs for 1 hour at 37°C.
- After washing the PO labeled goat-anti-horse IgG (commercially available
  from Cappel, Malvern, PH, USA) in a working dilution of 1 : 14000 in
  0.1 M PBS with sheep serum diluted 1 : 40, pH 7.2 and 0.05 % Tween 20
  is added and incubated for 1 hour at 37°C.
- After washing the substrate is added: 80 mg 5-amino-2-hydroxy-benzoic
  acid solved in 100 ml destilled water together with 0.05 % H_2O_2 in a 9 : 1
  ratio (for details see van Knapen et al., 1986).

ELISA antibody profile-expressed as reciprocal titres-of an experimentally with T.spiralis infected horse

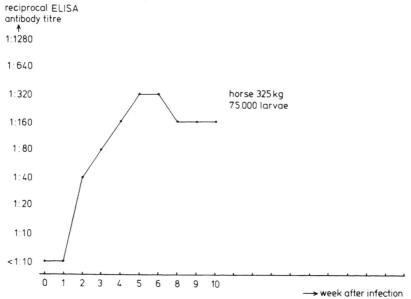

FIGURE 3

ELISA antibody profile-expressed as reciprocal titres-of an experimentally with T.spiralis infected pony

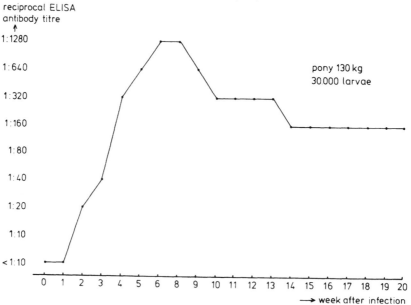

FIGURE 4

4. DISCUSSION AND CONCLUSIONS

In the past decade it has become obvious that horse meat is of potential danger to man with respect to T.spiralis infection when consumed rare. In fact, as many human victims have been reported due to consumption of infected horse meat as due to infected porc or wild boar meat all together in recent years in Europe.

Although herbivorous animals like horses normally are not expected to become infected by carnivorism, reality forces to reconsider this seriously. Particularly in countries with a tradition to consume rare or undercooked horse meat (Italy, France, Belgium) veterinary authorities should consider the introduction of compulsary meat inspection for T.spiralis in horses. Very little is known about horse T.spiralis infections however, large scale investigations with an artificial digestion method in Italy and France have never revealed a positive case (2, 5). Experimental infection of horses was carried out in Italy after the first outbreak ten years ago (9) and in Germany (14). Extreme variation in degree of infection between the animals was observed in both studies. The goal of our study was to set up a reliable serological method to demonstrate specific anti-T.spiralis antibodies. A horse and a pony were infected to obtain a substantial quantity of reference serum and to follow the immune response. From this study no conclusions can be made with regard to horses in general infected with T.spiralis, but the data obtained from the two animals are worth to be reported. First, surprisingly, no problems were observed to feed the animals with a skinned mouse. Although the infective dosage was relatively high (230 larvae/kg body weight) no clincial signs at all were observed. The T.spiralis strain used, however, was maintained in mice in the laboratory since 25 years and might be not very virulent for horse. Nevertheless, the individual differences are remarkable with regard to degree of muscle invasion. This was also observed in the Italian experiment (9). It is clear that horses may become easily infected with T.spiralis at a degree that may have serious public health consequenses when the meat is consumed rare. How long the trichinella larvae encapsulated in horse musculature remain viable warrants more research. From this experiment it may be suggested that, if artificial digestion of predilection sides as an inspection measurement is introduced preferably masseter musculature, tongue or diaphragm should be used. This is also in agreement with the italien (9) and german (14) experiences.

Introduction of a reliable serological assay for rapid screening of horses at a large scale would definitely contribute to collect more epidemiological information on horse trichinellosis. In this study we were able to set up an ELISA method which easily detects antibodies to T.spiralis after experimental infection. Discrimination between negative and positive sera was without problems due to very low background reaction in negative (preinfection) sera. In 157 sera tested from horses with different disorders, including a variety of parasitic infections, or horses without known history but originating trichinella endemic areas (Poland), no reaction in the ELISA method was seen. In one pony however, (Table 1) a strong reaction was observed. Unfortunately this animal could not be tested for T.spiralis because it was sacrificed long ago. Interestingly this animal was known to be infected with C.imparidentatum. It is doubtful whether this particular infection could have been the reason for a strong serological anti-T.spiralis immune reaction. It can not be excluded therefore that the animal indeed was infected with T.spiralis. Cross reaction due to the frequently occurring horse Strongylus species could be excluded by using Strongylus antigens in ELISA during the T.spiralis experimental infection. Both experimentally infected animals strongly reacted with specific antibodies although the muscle invasion was low in one and high in the other. Many questions still have to be answered: what will happen in low dosage infection, how long will antibodies remain in circulation, how long will encapsulated larvae remain viable, what about individual immune response in horses in general in case of Trichinella infection, can differences in background ELISA values be expected in horses form endemic and non-endemic areas as is reported from swine (6).

Some conclusions can be drawn so far:
- horses may easily become infected through rodents
- relatively high infection dosages are not a guarantee for clinical symptoms in horses
- ELISA is a suitable method to demonstrate antibodies to T.spiralis infection in experimentally infected horses.
- Predelection sites for T.spiralis in horses are M.masseter, tongue and diaphragm.

5. REFERENCES

1. Ancelle T, Dupouy-Camet J, Heyer F, Faurant C, Lappiere J. Outbreak of Trichinosis due to horse meat in the Paris area. The Lancet, 21 sept., 660, 1985.

2. Baldelli R, Morselli A e Stanzani F. Richerche sulla trichinellosi in cavalli macellati a Bologna. Parassitologia XXII (1-2) 157-160, 1980.

3. Bellani L, Mantovani A, Pampiglione S, Filippini I. Observations on an outbreak of human trichinellosis in Northern Italy. In: Trichinellosis, Proc.4th Int.Conf. on Trichinellosis. Aug.26-28, 1976, Poznan, Poland. Kim CW and Pawlowski ZS (eds). pp535-539. 1977.

4. Bouree P, Kouchner G, Gascon A, Fruchter J, Passeron J et Bouvier JB. Trichinose: bilan de l'épidémie de janvier 1976 dans lan banlieue sud de Paris (à propos de 125 cas). Ann.Med.Interne, 128, (8-9) 645-654, 1977.

5. Dupouy-Camet J. Personal communication. 1986.

6. Knapen F van, Franchimont JH, and Ruitenberg EJ (on behalf of the members of the EEC Working Group on Trichinellosis. The reliability of the Enzyme Linked Immunosorbent Assay (ELISA) for the detection of Swine trichinellosis. In: Trichinellosis, Proc.5th Int.Conf. on Trichinellosis. Sept.1-5, 1980, Noordwijk aan Zee, The Netherlands. Kim CW and Pawlowski ZS (eds). pp399-404, 1980.

7. Knapen F van, Buijs J and Ruitenberg EJ. Trichinella spiralis Antibodies. In: Methods of Enzymatic Analysis. 3rd ed. (Eds. Bergmeyer HU, Bergmeyer J and Grassl M. VHC mbH Weinheim, FRG. pp393-406, 1986.

8. Kohler G und Pfeiffer G. Zur Möglichkeit weiterer Verkürzung des direkten Trichinellennachweises beim Schlachtschwein. Fleischwirtsch. 63 (3) 330-331, 1983.

9. Pampiglione S, Baldelli R, Corsini C, Mari S e Mantovani A. Infezione sperimentale del cavallo con larve di trichina. Parassitologia XX (1,2,3), 183-191, 1978.

10. Ruitenberg EJ, Knapen F van. Enzyme linked immunosorbent assay (ELISA) as a diagnostic method for Trichinella spiralis infections in pigs. Vet.Parasitol. 3, 317-326, 1977.

11. Stein HA. Trichinose-Erkrankungen im Bitburger Raum (Eifel) aus der Sicht der Humanmediziner des öffentlichen Gesundheitsdienstes. Off.Gesundh.Wes. 45, 532-533. 1983.

12. Stumpf J, Kaduk B, Undeutsch K, Landgraf H und Gofferje H. Trichinose. Dtsch.Med.Wochenschr. 103 (40), 1-5, 1978.

13. Thomsen DU. Stomacher trikinkontrol-metoden. Dansk.Vet.Tidsskr. <u>59</u>, (1/6) 481-490, 1976.
14. Wöhrl H, Hörchner F und Grelck H. Zur Trichinellose des Pferdes. Rundschau für Fleischuntersuchungs und Lebensmittelüberwachung <u>30</u>, (3) 40-41, 1978.

INTESTINAL MAST CELLS: POSSIBLE REGULATION AND THEIR FUNCTION IN THE GUT OF
TRICHINELLA SPIRALIS INFECTED SMALL RODENTS

J.BUIJS[1], H.K.PARMENTIER[2], H.VAN LOVEREN[1] and E.J.RUITENBERG[1,2]

ABSTRACT
Mast cells were grown in vitro from bone marrow cells from normal and
infected rats in the presence of mast cell growth factor (MCGF) prepared
from different lymphoid organs at various times post infection (p.i.). It
was found that MCGF production by mesenteric lymph node cells (MLNC) in
vitro corresponded with the intestinal mast cell (IMC) response in vivo
during a T.spiralis infection, i.e. production of MCGF could be noted at 11
but not at 28 days p.i.. In contrast, spleen cells from day 28 p.i. still
produced MCGF in vitro. This is long after adult worms were expelled from
the gut. These data suggested that MCGF producing T cells in the spleen
remained activated by products from muscle larvae and induced the
maturation of connective tissue mast cells found near encapsulating larvae.
The type of tissue in which mast cells mature determined the resulting mast
cell subpopulation.
   By blocking serotonin receptors very early after infection it was
possible to suppress the inflammatory response in the gut. This suggested
that serotonin, probably by increasing vascular permeability, facilitated
entry of MCGF producing T cells or mast cell precursors in the gut mucosa.
Since no effect of serotonin blockers was measured on worm expulsion, these
results may indicate that mucosal mast cells play a minor role in the
expulsion of adult worms from the intestines.

1. INTRODUCTION
   Infection with the nematode Trichinella spiralis is associated with an
inflammatory response in the gut. This response, which is temporary, is
characterized by in influx of inflammatory cells with intestinal mast cells
(IMC) and eosinophils dominating. Other signs of inflammation are oedema of
the mucosa, submucosa and the muscle layers, shortening of the villi and an
increase in the number of goblet cells. The adult worms are expelled from
the gut in 14 (mice) or 18 (rats) days post infection (p.i.) after which
the inflammation subsides to normal at about 28 days p.i. The increase in
eosinophils is a reflection of increased proliferation of eosinophil
precursors in the bone marrow (BM). Mast cell precursors, however, were
never recognised as such. Schrader et al. (19) succeeded in growing mast
cells from mouse bone marrow cells (BMC) showing that bone marrow contained
mast cell precursors. The mast cell growth stimulating agent (MCGF), now
known as interleukin-3 (IL-3) (20), consisted of culture supernatant
prepared from Concanavalin A stimulated normal mouse spleen cells. Haig et
al (6, 7) using the Nippostrongylus brasiliensis rat model, prepared
culture supernatant from specifically stimulated lymph node cells (MLNC).
This supernatant induced the growth of mucosal mast cells when added to BMC

cultures. The cultured mast cells contained a lower sulphated glucosaminoproteoglycan instead of heparin (15), i.e. typical for IMC.

To study IMC regulation we investigated in the T.spiralis/rat model the MCGF activity in conditioned media (CM). CM were prepared from different lymphoid organs at various times p.i.. Since IMC responses coincided with the gut phase of T.spiralis infection it was suggested that IMC were directly involved in the expulsion of adult worms. Therefore the times chosen to harvest the lymphoid tissues were just before (11 days p.i.) and after (28 days p.i.) highest IMC counts.

Furthermore, the relation of elevated IMC numbers with adult worm expulsion was investigated. It was proposed that, analogous to delayed type hypersensitivity (DTH) reactions, the increase in inflammatory cells in the gut in response to a T.spiralis infection depended on (T-cell dependent) serotonin release by serotonin containing cells, e.g. resident gut mast cells (14). Serotonin induces vasopermeability by binding to serotonin receptors on the endothelial cells of neighbouring blood capillaries. This allows the sequential influx of immune and inflammatory cell populations. If the inflammatory response in the gut is dependent on serotonin, it should be possible to competitively block the serotonin receptors with a serotonin antagonist (10). It was shown that by treating mice with methysergide (the amino acid alkaloid ergotamine), the serotonin receptors were blocked, resulting in suppression of the inflammatory response in the skin (10). This paper summarizes the results of experiments mentioned above and discusses the role of the MLN in IMC regulation and the effect of interference with IMC numbers on adult worm expulsion.

2. MATERIALS AND METHODS

2.1. Animals and parasites:

Inbred female thymus bearing rats (Wag/Cpb +/rnu) and inbred female thymus bearing mice (B10LP +/nu) 6-10 weeks of age were used. The T.spiralis strain originated from a Polish boar and is maintained in small rodents. Infectious muscle larvae were obtained by digestion of rodent carcasses in HCl and pepsine solution (8). The infection was given orally (2500 muscle larvae/rat and 300 muscle larvae/mouse).

2.2. Preparation of MCGF-containing conditioned medium (CM)

Rats were infected with T.spiralis muscle larvae. At days 11 and 28 p.i. MLN, spleen, mandibular lymph nodes were removed. Single cell suspensions were prepared in Gibco's RPMI-1640 + L-Glutamine supplemented with 5% foetal calf serum (FCS), 10 IU heparin/ml, 3 mg DNase/100 ml, 200 IU/ml of penicillin and 0,2 mg streptomycin/ml. The cells were seeded in tissue culture flasks at a density of 10^6/ml. T.spiralis antigen (10 µg/10^6 cells) or Con A (0,31 µg/10^6 cells) was added. The cultures were incubated at 37°C, in humidified air containing 5% CO_2 for 3-4 days after which the supernatants (= CM) were collected. A similar procedure was followed for the preparation of CM from MLN and spleen cells from normal rats and from rats given a booster infection 21 days after primary infection.

2.3. Mast cell cultures

BMC were obtained by flushing the tibia and femur from infected and normal rats. Single cell suspensions were prepared in RPMI-1640 + L-Glutamin plus supplements (20% FCS). The cells were seeded in 6 well culture plates (Costar, Cambridge, Mass, USA) at a density of 0,75 x 10^6/ml. The BMC were grown in above mentioned RPMI-1640 supplemented with 50% CM. The cultures were incubated at 37°C in 5% CO_2 in humidified air. Total cell numbers and/or relative mast cell numbers were estimated at various days post incubation. Total cell counts were done using a Bürker haemocytometer. Relative mast cell and eosinophil numbers were estimated by counting their number per 300 cells in May–Grünwald/Giemsa stained preparations.

2.4. Treatment with serotonin (receptor) antagonist

Groups of 5 mice were orally infected with 300 T.spiralis muscle larvae at day 0. Methysergide (Sandoz, Pharmaceuticals, Basel, Switserland), a competitive antagonist of serotonin, 1 mg in PBS/animal was inoculated subcutaneously into mice (80 mg/kg/day) twice a day at various days in between day −1 and day +13 p.i. The dose of 1 mg methysergide was found to inhibit ear swelling induced by inoculation of 15 µg serotonin. Groups of mice were sacrificed 14 days p.i. Swiss rolls were made of the jejunum, fixed in formalin (10%) and in formalin/acetic acid (0,8% / 4%) (18). After embedding in paraffin 5 µm sections were stained with haematoxylin and eosin or with Toluidin Blue. IMC, eosinophils and goblet cells were counted in 20 villus/crypt units per animal. Worm expulsion from the gut was determined in mice orally infected with 300 T.spiralis muscle larvae and in infected mice treated twice daily with methysergide from day −1 until day +8. Groups of mice were killed at days 10, 12 and 14 p.i. Counting of adult worms was done as described previously (16).

3. RESULTS

3.1. Mast cell cultures

CM prepared from normal mandibular lymph nodes (ManLN) at 11 and 28 days p.i. were added to BMC from normal and infected rats (table 1). CM prepared from normal ManLN contained very low MCGF activity. CM prepared at 11 and 28 days p.i. and stimulated with the specific antigen also showed very low MCGF activity. However, when stimulated with Con A especially the CM from 11 days p.i. had high MCGF activity, many mast cells developed from the BMC. No differences were observed in the results using BMC from normal and infected rats.

Table 1

CM were prepared from ManLN from normal and infected rats (11 and 28 days p.i.). the cells were stimulated with T.spiralis antigen (Ag) or with Con A and were added to BMC from normal and infected (11 days p.i.) rats. The relative mast cell numbers were estimated by counting the mast cells/300 cells in total in May-Grünwald– Giemsa stained preparations

CM		Days Postincubation	3		10		16	
	BM		norm.	infect.	norm.	infect.	norm.	infect.
ManLN normal/Ag			ND	0	ND	0	ND	ND*
/ConA			ND	0,7	ND	3	ND	0
11	/Ag		1	2,7	3,3	4	3,3	4,7
11	/ConA		28,3	11,7	42,7	31,7	60,3	63,3
28	/medium (+)		ND	0	ND	0	ND	ND*
28	/Ag		0	0,7	0	2	ND*	ND*
28	/ConA		5,7	11	12	4,7	13,3	ND*

ND = not done
ND* = number of cells per culture was very low
(+) = cultured without stimulants

The results shown in the Figures 1 – 5 were obtained by culturing BMC from normal and infected (11 days p.i.) rats in the presence of CM prepared from MLNC and spleen cells. CM prepared from antigen stimulated MLNC at 11 days p.i. had high MCGF activity, about 80% of the total cell population consisted of mast cells (figure 1). The Con A stimulated MLNC from day 11 p.i. produced also MCGF with high activity; 74% of the cells being mast cells at day 13 of culture. CM prepared from day 28 MLNC did not have the property to produce MCGF. Both CM prepared from antigen and Con A stimulated cells induced very low numbers of mast cells in the culture (Figures 1 and 2). The cultures in which BMC from infected rats were used showed high relative numbers of eosinophils. The cultures using BMC from infected rats and CM from antigen stimulated MLNC (11 and 28 days p.i.) were repeated. The absolute numbers were estimated (Figure 3). Using day 11 p.i. CM the number of mast cells increased very rapidly reaching peak values at day 9 of culture which was maintained until the end of the observation period (day 18). The total number of cells decreased to approximately 30% at day 9 and to approximately 10% of the initial population at day 18 of culture. CM from day 28 showed low MCGF activity ($\geq$ 1 %) and induced rapid decrease in total cell number from day 9 and onwards, reaching 1% of the initial population at day 18. The number of eosinophils remained at starting values until day 9 after which it decreased parallel to that of the total cell number.
Spleen cells stimulated with either antigen or Con A and collected at 11 and 28 days p.i. produced high MCGF activity. (Figures 4 and 5). In all cultures the mast cell number was 40 – 60% of the total cell population at day 13 of culture. Using CM from Con A stimulated spleen cells harvested at day 11 p.i. the relative mast cell number was 53%. All conditioned media prepared from MLN and spleen cells were assayed on BMC from normal and

Course of relative numbers of mast cells and eosinophils

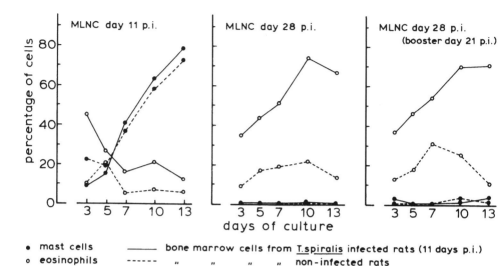

● mast cells   ———— bone marrow cells from <u>T.spiralis</u> infected rats (11 days p.i.)
○ eosinophils   ------   "   "   "   "   non-infected rats

FIGURE 1   Bone marrow cells cultured in the presence of conditioned medium prepared of MLNC from T.spiralis infected rats and stimulated with specific antigen

Course of relative numbers of mast cells and eosinophils

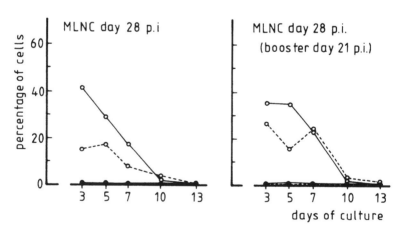

● mast cells;   ○ eosinophils;   —— bone marrow cells from infected rats (11 days p.i.);   ---- from non-infected rats

FIGURE 2   Bone marrow cells cultured in the presence of conditioned medium prepared of MLNC from T.spiralis infected rats and stimulated with Con-A

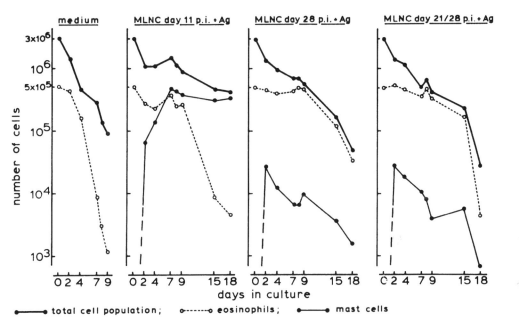

FIGURE 3 Bone marrow cells from infected rats cultured in the presence of conditioned medium prepared of MLNC from T.spiralis infected rats and stimulated with specific antigen. Counts of absolute numbers of mast cells and eosinophils

Course of relative numbers of mast cells and eosinophils

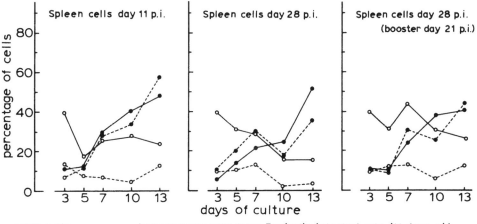

FIGURE 4 Bone marrow cells cultured in the presence of conditioned medium prepared of spleen cells from T.spiralis infected rats and stimulated with specific antigen

208

Course of relative numbers of mast cells and eosinophils

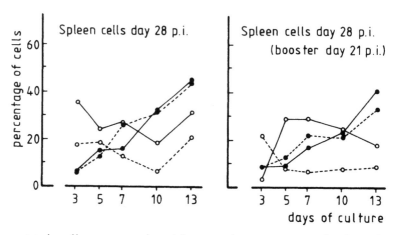

• mast cells;   ○ eosinophils;   —— bone marrow cells from infected rats (11 days p.i.);   ---- from non-infected rats

FIGURE 5   Bone marrow cells cultured in the presence of conditioned medium
prepared of spleen cells from T.spiralis infected rats and stimulated with
Con A

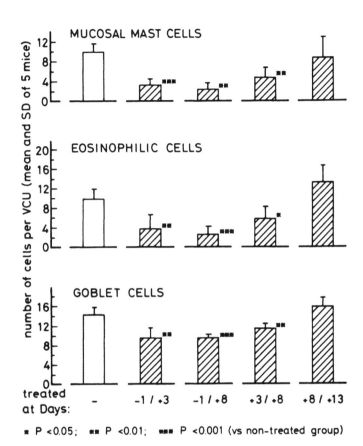

* P <0.05; ** P <0.01; *** P <0.001 (vs non-treated group)

FIGURE 6 Effect of Methysergide (twice daily 1 mg IM) on mucosal mast cells, eosinophilic cells and goblet cells in jejunum of B10LP +/nu mice at Day 14 after oral infection with 300 T.spiralis larvae

infected rats (day 11 p.i.). No differences were found.

3.2. Methysergide treatment

Infected mice were treated with the serotonin antagonist methysergide at days −1 till +3, −1 till +8, +3 till +8 or +8 till +13 p.i. Mucosal mast cells (IMC), eosinophils and goblet cells were counted in the jejunum at day 14 p.i. The results are given in Figure 6. Treatment with methysergide significantly depressed the mast cell, eosinophil and goblet cell responses to infection. The effect of methysergide treatment on worm expulsion is presented in table 2. Groups of mice were treated from day −1 until day +8 p.i. The number of adult worms present in the gut were estimated at days 10, 12 and 14 p.i. The number of worms isolated from the Methysergide treated animals did not significantly differ from the number isolated from the non-treated animals.

Table 2

Groups of mice were orally infected with 300 T.spiralis muscle larvae each. From day −1 until day +8 the animals were subcutaneously inoculated twice daily with 1 mg methysergide/animal. The number of worms present in the jejunum were estimated at 10, 12 and 14 days p.i.

	Number of worms isolated from the jejunum		
	day 10 p.i.	day 12 p.i.	day 14 p.i.
Methysergide treated	73,2 + 13,5	26,4 + 18,1	13,6 + 6,9
Control	73,0 + 49,1	35,6 + 24,2	14,6 + 10,1

* results are the means + sd of 5 - 8 animals

4. DISCUSSION

The rapid increase in IMC in the gut is followed by an equally rapid decrease and parallels the intestinal phase of a T.spiralis infection. This T cell dependent process lasts 3-4 weeks (17). The aim of this investigation was to study the site of IMC regulation and the IMC involvement in adult worm expulsion from the gut. IMC regulation was investigated by assaying MCGF production in vitro by T cells from different lymphoid organs at various times p.i. The MCGF target population were the mast cell precursors in the bone marrow. The activity of MCGF produced in vitro by MLNC at 11 and 28 days p.i. reflected the IMC response in vivo. MCGF activity in the CM prepared at 11 days p.i. induced elevated numbers of mast cells in the BMC cultures (Figures 1, 3). In CM prepared at 28 days p.i. the MCGF activity was very low and only few mast cells were present in the cultures (Figures 1, 3). In the in vivo situation IMC number is increasing at 11 days p.i. and has returned to almost control level at 28 days p.i. No differences in MCGF activity were observed after activation with either antigen or Con A (Figure 2). In contrast to the observation of Haig et al. (6, 7) who used N.brasiliensis as infectious agent, we did not find at any occasion that more mast cells were cultured from T.spiralis infected BMC than from BMC from non-infected rats.

Parmentier et al (15) estimated by a limiting dilution method the number of mast cell colony forming units in the BM of normal and T.spiralis infected mice. These authors did not observe any difference either. Using CM from ManLNC it was found that no antigen specific MCGF producing T cells were present. Only after Con A stimulation CM prepared from day 11 p.i. cells had good MCGF activity. MCGF activity in CM prepared from spleen cells did not show any variation. All BMC cultures to which CM was added prepared from antigen or Con A stimulated cells from 11 and 28 days p.i. had relative mast cell numbers of 40% to 60%. However, MCGF activity in CM from Con A stimulated spleen cells from normal rats induced already mast cell growth reaching values of 30% of the total cell population against less than 5% in Con A stimulated normal MLNC (results not shown). This shows that the spleen normally already possesses a rather large MCGF producing T cell population.

The following questions arose from these data 1) why did the MCGF producing T cell population in the MLN decrease so rapidly and 2) why was the antigen specific MCGF producing T cell population in the spleen not reflected by an IMC response in the gut (i.e. at 28 days p.i.).

1) Congenitally T cell-deficient small rodents infected with T.spiralis do not respond with an increase in IMC (16), although IMC precursors are present in the gut (15). Transfer of T.spiralis immune cells from thymus-bearing rodents to normal or congenitally thymus deficient animals does not result in IMC response in the gut either, unless the adult worms are present. So both antigen and functional T cells are required. Apparently, MLN from normal non-infected animals contain very few MCGF producing T cells (data not shown). Upon antigen stimulation this number increases and once the stimulus is gone decreases to control level. This explains why the increase and decrease in MCGF producing T cells follow so closely the entrance in and exit of adult worms from the gut. The MCGF producing T cells react to a local infection returning to normal after its removal. This latter phenomenon explains why ManLN from infected rats fail to produce MCGF after antigen stimulation, although they can do so after Con A stimulation, i.e. they are not specifically stimulated during infection.

Challenge infection given at 21 days p.i. was not reflected in a further increase in MCGF producing T cells seven days later (Figures 1, 2, 3, 4). Also in earlier experiments it was found that in the in vivo situation a challenge infection given at 14 days p.i. did not booster the IMC number in the gut. This is explained by the fact that worm expulsion was completed within 7 days after challenge infection (18), and antigen had therefore vanished by day 28, resulting in a decrease in MCGF production. Lee et al found that the worms were trapped in the mucous layer still present on the gut lumen surface and did probably not enter the mucosa (9). This corroborates our hypothesis that the presence of antigen is required for MCGF production in vivo.

2) In contrast to MLN, the spleen contains T cells that are able to produce MCGF when challenged with antigen in vitro 28 days after infection of the donor rats. We explain this by assuming that, in contrast to lymphoid cells of intestine-draining MLN, lymphoid cells in other lymph nodes, and consequently lymphocytes in the spleen are still being stimulated by another source of T.spiralis antigens then adult worms in the gut, i.e. the muscle larvae of T.spiralis, that by that time are abundantly available. During encapsulation of muscle larvae antigens are still leaking. A major function of the spleen is phagocytizing foreign and degenerating products. Worm antigen will be present in the spleen as long as it is in circualtion. Thus, MCGF producing T cells therefore remain

activated.

Connective tissue mast cells (CTMC) are prominent cells present in the cellular inflammatory reaction surrounding encapsulating larvae until at least 42 days p.i. (4) The maturation of these CTMC may be stimulated by spleen cell (or other lymphoid organ) derived MCGF. Nakano et al (12) inoculated cultured mast cells with MMC characteristics into the peritoneal cavity of $WBB6F_1$, $-/WW^V$ mice. 10 Weeks later various tissues were searched for the presence of mast cells. In the peritoneal cavity, spleen, skin and muscularis propria, mast cells were found which stained with berberine sulfate and safranin, both typical for CTMC. However, in the gastric mucosa mast cells were found which stained only with alcian blue. These results suggest that the origin of MCGF is not relevant in determining the type of mast cell but rather the tissue in which the cells mature, or CTMC require, besides MCGF, additional factors, either derived from the tissues, or the spleen, as has been proposed recently (13). Ginsburg et al (3) mentioned that growth of CTMC in vitro required the presence of fibroblasts. Also Czarnetzky et al. (2) succeeded in growing CTMC in the presence of transformed mouse fibroblast (L cells) culture supernatant.

The reason that the antigen sensitive MCGF producing T cells in the spleen do not induce IMC in the gut is that the worms have been expelled and thus the antigen stimulus attracting MCGF producing T cells to the gut.

Our results indicate that serotonin, probably by allowing mast cell precursors, or MCGF producing T cells, to leave the circulation and enter the extravascular tissue, is involved in the mucosal mast cell response to infection with T.spiralis. Treatment of infected mice with methysergide showed that the immune response starts very early after antigen entrance. Blocking the serotonin receptors from day -1 until day 3 p.i. already suppressed the IMC response (Figure 7), as was true if treatment was continued until day 8 p.i. Treatment from day 8 p.i. and onwards remained ineffective showing that the immune response may be comprised of a series of sequential reactions. Treatment with methysergide, that blocked the development of IMC response, did not interfere with expulsion capacity of the infected mice. This suggests that the mucosal mast cells that arrive on the scene during a T.spiralis infection play only a minor role in expulsion of the adult worms. Mitchell et al (11), using mast cell deficient W/W^V mice, showed that expulsion of N.brasiliensis was only slightly delayed. Ha et al (5), also using W/W^V mice, found that expulsion of T.spiralis was somewhat delayed but certainly not inhibited. However, Alizadeh and Wakelin (1) found that low IMC responders were also slow responders with regard to expulsion suggesting a causal relationship. The function of IMC in the gut inflammatory response, therefore still remains to be elucidated.

5. REFERENCES

1. Alizadeh H and Wakelin D. Genetic factors controlling the intestinal mast cell response in mice infected with Trichinella spiralis. Clin.Exp.Immunol. 49, 331-337, 1982.
2. Czarnetzki BM and Behrendt U. Studies on the in vitro development of rat peritoneal mast cells. Immunobiol. 159, 256-268, 1981.
3. Ginsburg U, Ben-Shahar D and Ben-David E. Mast cell growth on fibroblast monolayers: two-cell entities. Immunol. 45, 371-380, 1982.
4. Gustowska L, Ruitenberg EJ and Elgersma A. Cellular reactions in tongue and gut in murine trichinellosis and their thymus-dependence. Parasite Immunology 2, 133-154, 1980.
5. Ha TY, Reed ND and Crowle PK. Delayed expulsion of adult Trichinella spiralis by mast cell-deficient W/W^v mice. Infect. Immun. 41, 445-447, 1983.
6. Haig DM, McKie TA, Jarrett EEE, Woodbury R and Miller HRP. Generation of mucosal mast cells is stimulated in vitro by factors derived from T cells of helminth infected rats. Nature, 300, 188. 1982.
7. Haig DM, Jarrett EEE and Tas J. In vitro studies on mast cell proliferation in N.brasiliensis infection. Immunology 51, 643-651. 1984.
8. Köhler G and Ruitenberg EJ. Comparison of three methods for the detection of Trichinella spiralis infections in pigs by five European laboratories. Bull. Wld. Hlth. Org. 59, 413-419. 1974.
9. Lee SB and Ogilvie BM. The mucus layer of the small intestine. Its protectie effect in rats immune to Trichinella spiralis. In: Trichinellosis. ICT-5. Kim WC, Ruitenberg EJ and Teppema JS (eds). Reedbooks Ltd. Chertsey Surrey, England, pp91-95, 1980.
10. Loveren H van, Kraenter-Kops S and Askenase PW. Different mechanisms of release of vaso-active amines by mast cells occurr in T-cell-dependent compared to IgE-dependent cutaneous hypersensitivity responses. Eur.J.Immunol 14, 40-47, 1984.
11. Mitchell LA, Wescott RB and Perryman LE. Kinetics of expulsion of the nematode Nippostrongylus brasiliensis in mast cell deficient W/W^v mice. Parasite Immunol. 4, 1-12, 1983.
12. Nakano T, Sonoda T, Hayashi C, Yamatodani A, Kanayama Y. Ymamura T, Asai H, Yonezawa T, Kitamura Y and Galli SJ. Fate of bone marrow-derived cultured mast cells after intracutaneous intraperitoneal and intravenous transfer into genetically mast cell-deficient W/W^v mice. J.Exp.Med. 162, 1025-1043, 1985.
13. Nakahata T, Kobayashi T, Ishiguro A, Tsuji K, Naganuma K, Ando O, Yagi Y, Tadokoro K, Akabane T. Extensive proliferation of mature connective tissue-type mast cells in vitro. Nature 324, 65-67, 1986.
14. Parmentier HK, Vries C.de, Ruitenberg EJ and Loveren H van. Involvement of serotonin in intestinal mastocytopoiesis and inflammation during a Trichinella spiralis infection in mice. Int.Archs.Allergy Appl.Immun. (in press).
15. Parmentier HK, Teppema JS, Loveren H van, Tas J and Ruitenberg EJ. Effect of a Trichinella spiralis infection on the distribution of mast cell precursors in tissues of thymus bearing and non thymus bearing (nude) mice determined by an in vitro assay. Submitted.
16. Ruitenberg EJ and Steerenberg PA. Intestinal phase of Trichinella spiralis in congenitally athymic (nude) mice. J.Parasit. 60, 1056-1057. 1974.
17. Ruitenberg EJ, Elgersma A. and Kruizinga W. Absence of intestinal mast

214

cell response in congenitally athymic mice during Trichinella spiralis infection. Nature 264, 258-260, 1976.

18. Ruitenberg EJ and Elgersma A. Study on the kinetics of globule leucocytes in the intestinal epithelium of rats after single or double infections with Trichinella spiralis. Br.J.Exp.path. 61, 285-290, 1980.

19. Schrader JW, Lewis SJ, Clark-Lewis J and Culvenor JG The persisting (P) cell, histamine content regulation by a T-cell derived factor, origin from a bone marrow precursor and relationship to mast cells. Proc.Natl.Acad.Sci. 78, 323-327, 1981.

20. Yung UP, Eger R, Tertian G and Moore MAS. Long-term in vitro culture of murine mast cells. II. Purification of a mast cell growth factor and its dissociation from TCGF. J.Immu. 127, 794-799, 1981.

VARIABLE LEVELS OF HOST IMMUNOGLOBULIN ON MICROFILARIAE OF BRUGIA PAHANGI ISOLATED FROM THE BLOOD OF CATS

U.N. Premaratne , R.M.E. Parkhouse and D.A. Denham

1. ABSTRACT

Using a variety of polyclonal affinity purified antibodies and monoclonal antibodies against humoral components of cat serum we have demonstrated cat IgG on the microfilarial sheath of Brugia pahangi isolated from the blood of cats. No feline IgG could be seen on microfilariae born to pregnant adult female worms maintained in vitro, nor on the surface of microfilariae whose sheaths had been removed. Microfilariae are born without IgG on their sheaths and slowly accumulate it as they circulate in the blood. This process results in variable levels of host immunoglobulin on the microfilariae. It is not known whether the IgG protects the worm from recognition by the host or that it eventually accumulates to such a level that it mediates killing of the worm by granulocytes.

2. INTRODUCTION

Many helminths survive within their hosts despite the presence of host immunoglobulin (Ig) on their surfaces. On the other hand, antibody bound to several nematodes may cause adherence of granulocytes in vitro, with consequent damage to the cuticle, as for example with newborn larvae of Trichinella spiralis (15, 13) or with microfilariae of Dipetalonema viteae, (22, 23) Litomosoides carinii, (21) and Brugia pahangi (11). Not all antibody classes or isotypes, of course, necessarily mediate these granulocyte killing mechanisms, and indeed those that do not may act as blocking antibodies (9). Whatever the outcome in individual cases, the presence or absence of host Ig on parasite surfaces may be a decisive factor in parasite survival. We have, therefore, reinvestigated (20) the question of whether there is host Ig on the surface of microfilariae of B. pahangi, using a combination of affinity purified and monoclonal anti-cat Ig reagents.

The present study showed that host immunoglobulin is acquired by microfilariae of B. pahangi after their birth and during their circulation in the blood of cats.

216

3. MATERIALS AND METHODS

3.1 Cats

The cats (Felis catus) used were laboratory bred and
maintained in open rooms with full ethical support.

32. Parasites

B. pahangi was maintained in cats using Aedes aegypti as
the vector.   Procedures for infecting cats were described
(4).   The microfilariae were separated from venous blood
using a Nuclepore 5um pore filter (Dennis and Kean 1977)
and, whilst retained on the membrane in the filter holder,
were washed twice in sterile phosphate buffered saline
(PBS), once with distilled water, once with sterile PBS,
once with RPMI-1640, and resuspended in RPMI-1640.

To obtain microfilariae which had not been exposed to
feline Ig adult worms were recovered from the peritoneal
cavities of jirds (Meriones unguiculatus), which had been
inoculated intra- peritoneally with infective larvae three
months earlier.   The adults were incubated for 3 hours in
vitro in 10% (v/v) foetal calf serum (FCS) in RPMl-1640
and the microfilariae born collected.

3.3 Preparation of cat IgG and F(ab')2

Cat IgG was prepared from a pool of normal cat serum by
adsorption onto protein-A-sepharose (Pharmacia); it was
eluted with 0.2M glycine-HCl at pH 2.8, neutralised with
solid tris, and dialysed against 0.13M NaCl.   Cat IgG
was converted to F(ab')2 by digestion with pepsin (pepsin
2% w/w of IgG, 0.2M NaAc for 16 hrs, 37oC at pH 4.2).

3.4 Preparation of affinity purified rabbit anticat IgG antibodies

Rabbits were immunized with cat IgG as described (18).
Cat IgG was coupled to cyanogen bromide activated
Sepharose 4B (Pharmacia) (1).   Rabbit anti-cat IgG
antiserum was passed through cat IgG-sepharose columns.
The anti-cat IgG antibody was eluted with 0.2M glycine-
HCl, pH 2.4,   neutralized with solid tris and dialysed
against 0.13M NaCl. To prepare anti-F(ab')2 antisera
rabbits were immunized with cat F(ab')2 and the
antibodies affinity purified by adsorption onto a column
of cat IgG-Sepharose followed by elution with 0.2M
glycine-HCl at pH 2.4. Anti-cat Fc(IgG) was collected in
the unbound fraction after the affinity purified rabbit
anti-cat Ig was passed through a column of cat F(ab')2-
sepharose.

Rabbit IgG antibodies were converted to F(ab')2 by pepsin

digestion. The affinity purified IgG and F(ab')2 reagents were conjugated with fluorescein (FITC) (17).

3.5 Monoclonal antibodies to cat IgG

Mouse monoclonal antibodies to cat mu chains (CC4), gamma chains (AC5), Kappa light chains (BA3) and Lambda light chains (EA2) were donated by Drs. F.Klotz and M.D.Cooper of The University of Alabama, Birmingham, Alabama, U.S.A.). Also mouse monoclonal antibodies were prepared by us against cat IgG and IgM by the technique of Kohler and Milstein (14).

3.6 Goat anti-mouse IgG

A goat anti-mouse Ig, reactive with all mouse Ig classes was affinity purified (3) and conjugated with FITC as described above.

3.7 Immunofluorescence tests

Isolated microfilariae were washed twice in PBS/bovine serum albumin(BSA)/sodium azide (BSA 2mg/ml, azide 2mg/ml), reacted with FITC rabbit F(ab')2 anti-cat F(ab')2 (50-100 ug/ml) for 20 min in Luckham tubes on ice, washed twice in PBS/BSA/azide and examined for fluorescence with a Zeiss incident light fluorescence microscope.

Positive fluorescence was considered complete staining with an uninterrupted line of yellow-green fluorescence around the sheath or the cuticle (in the case of exsheathed worms). Positive fluorescence was scored as 0 to 3+ depending on its subjectively assessed intensity.

3.7 Immune cat serum

On some occasions immune cat serum was used. This serum was taken from cats known to be resistant to challenge (5) and to stimulate lethal granulocyte adherence in vitro (11).

4. RESULTS

4.1 Demonstration of cat immunoglobulin on the sheath of microfilariae

When microfilariae freshly taken from the blood of cats were incubated with affinity purified FITC rabbit anti-cat F(ab')2 or anti-cat Fc' fluorescence was clearly seen on the sheath of most microfilariae. Subsequent investigation showed that microfilariae also reacted with mouse monoclonal antibodies to cat heavy (gamma) and light (kappa and lambda) chains. In these tests binding of mouse monoclonal antibody was revealed by a second layer of FITC goat anti-mouse Ig. Fluorescence was not observed on the sheaths of microfilariae exposed to normal cat Ig followed by FITC anti-cat Ig, FITC goat anti-mouse Ig, FITC normal rabbit IgG, mouse monoclonal antibodies to cat mu chains nor with an affinity purified goat anti-human alpha chain antibody, which cross reacts with cat IgA (Klotz, F. and Cooper, M.D., personal communication). Neither was fluorescence seen on microfilariae isolated from the blood of infected jirds or produced by adult B. pahangi maintained in vitro (and, thus, never exposed to feline antigens) incubated with the same anti-cat reagents as above.

Microfilariae collected from adult worms maintained in vitro were incubated for 20 minutes on ice with serum from cats known to be resistant to challenge with B. pahangi (5). Such sera contain IgG antibodies which have been shown to mediate cell adherence leading to microfilarial death (11). All of these microfilariae were shown to absorb IgG by their subsequent reaction with rabbit anti-cat F(ab')2.

In order to determine if host IgG was present on the cuticular surface of the microfilariae as well as on the sheath, microfilariae from an old B.pahangi infection were exsheathed in 10 mM calcium chloride (7). No fluorescence was seen on the cuticular surface of these exsheathed microfilariae after incubation with FITC rabbit anti-cat F(ab')2, but when they were pre-incubated with immune cat serum, followed by FITC rabbit anti-cat F(ab')2 they fluoresced very strongly.

4.2 The degree of fluorescence on individual microfil- filariae varies

Although we could only measure the degree of fluorescence subjectively, frequent checks were made by colleagues. In every case we noted that a proportion of the microfilariae examined did not appear to have Ig on their sheaths and of those with IgG on their sheaths, some fluoresced very strongly while others fluoresced less strongly. This is illustrated in Figure 1.

The proportion of microfilariae which fluoresced also varied from cat to cat. Usually in older infections with stable microfilaraemias about 80% of the microfilariae fluoresced.

This variation in fluorescence was seen whether anti-F(ab')2, anti-Fab, anti-Fc or anti-IgG reagents were used. Typically 10-15% of microfilariae were grade 3+, 25-50% 2+, 20-40% 1+ and 10-20% were negative. When reacted with the immune cat serum followed by FITC anti-cat Ig, however, cell microfilariae, whether uterine or circulating, were stained.

Previous studies have shown that young microfilariae differ from older microfilariae in their infectivity to mosquitoes (10) and in their ability to activate complement (11). This prompted us to investigate whether young microfilariae differed in the amount of IgG on their sheaths. To this end cats were inoculated with the infective larvae of B. pahangi and checked regularly to determine when they became microfilaraemic. Microfilariae were collected one and two months after their first appearance in the blood and also from cats which had been microfilaraemic for more than six months. Due to the scarcity of microfilariae in circulation for the first month of microfilaraemia it was impractical to collect sufficient to observe younger specimens.

Only about 10% of the microfilariae taken from cats which had been microfilaraemic for less than one month reacted with FITC in cat Ig reagents. This rose to about 25% of the microfilariae from the same cats one month later and to more than 75% of microfilariae after more than six months of microfilaraemia.

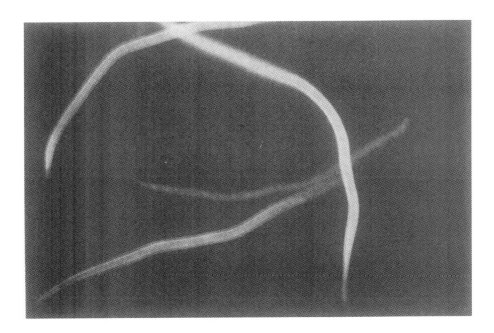

FIGURE 1. Photomicrographs of microfilariae of Brugia pahangi seen with incident UV light microscopy. The microfilariae were collected from cat blood, washed and reacted with FITC affinity chromatography purified rabbit F(ab')2 anti-cat F(ab')2.

4.3 Stability of surface bound host Ig to the microfilariae

About 1500 microfilariae collected from an old infection were added to 10 ml of medium (RPMI-1640, supplemented with 10% FCS, penicillin and streptomycin) in a sterile 10 ml screw capped tube, and incubated at 37oC in an atmosphere of 5% CO_2 in air. Daily, 9 ml of culture medium were removed and replaced with fresh medium. Samples of microfilariae were removed from the cultures on days 1, 2, 6 and 9 and examined for surface bound host immunoglobulin by staining with FITC F(Ab)2 anti-cat IgG. The number of stained microfilariae dropped from 87% at t=0 to 62% at 24 hr, and then slowly diminished to 47% on day 6. Whilst this indicates that the attachment of IgG to the sheath is not permanent a large part of the antibody is very firmly fixed. If the cultured microfilariae were subsequently incubated in immune cat serum all of them showed an antibody layer on their sheaths.

5. DISCUSSION

The principal observation reported in this paper is that there are variable amounts of entire IgG antibodies on the surface of the sheath of B. pahangi microfilariae recovered from the blood of cats. This was detected using affinity purified-rabbit anti-cat F(ab')2 and Fc reagents, and anti-gamma and light chain mouse monoclonal antibodies. We conclude that the bound cat IgG antibody was intact because antibodies directed against both Fab and Fc portions of the molecule produced similar positive reactions. No reaction was seen with either anti mu or alpha chain antisera, nor did normal cat IgG bind to the surface of microfilariae.. We also conclude that microfilariae are born without this coat of host Ig antibody as only a small proportion of microfilariae from cats which have only recently become microfilaraemic have antibody on their surface. As time progresses more microfilariae display host IgG on their surface. Even in cats which have been microfilaraemic for a year, or more, 15-20% of microfilariae still show no or only feeble fluorescence (Figure 1). The surface of Trichinella spiralis expresses protein molecules which change qualitatively after moulting and quantitatively during the growth of the worms within one life-cycle stage (12, 19). If there were a similar change in the surface of the sheath of microfilariae with age, this might explain the difference in the uptake of immunoglobulins onto their surface.

We have no evidence as to the functional significance of the IgG on the surface of the microfilariae. It could serve as a disguise on microfilariae but why the cat's granulocytes are not focussed via their Fc receptors in vivo as they are when microfilariae are exposed to immune cat serum in vitro (11) is perplexing. One explanation is that antibody is gradually absorbed onto the sheath and that only those microfilariae which have accumulated sufficient surface Ig are eliminated in vivo by such mechanisms. Given the variable amount of surface Ig described in this study, most microfilariae would not be enshrouded in such a level of bound host Ig. An alternative explanation is that during the microfilaraemic phase of the infection the anti-sheath antibodies produced do not present Fc receptors to the granulocytes or recognize, what to the parasites are, significant epitopes. Only after the cat has mounted an effective immune response (5) does cat serum contain antibody which takes part in a lethal event involving granulocytes in vivo. The surface of the sheath is known to present a limited but multiple number of antigens (16) and cats which become amicrofilaraemic and, therefore, resistant to reinfection recognize certain antigens not recognized by microfilaraemic cats as demonstrated by Western blotting techniques.

The masquerade of microfilariae behind a potentially lethal disguise may also benefit the parasite through immune elimination of senescent microfilariae which might otherwise compete with younger forms which might be more infective to mosquitoes. This could be a case of "having your cake and eating it". That only IgG could be detected on microfilariae does not necessarily mean that IgM and IgA are never involved in responses to the surface of microfilariae at any stage of their life cycle. It might be that IgM, for example, is present only very early in an infection possibly when microfilariae first start to circulate or just before the onset of a microfilaraemic state. Indeed control of microfilaraemia of D. viteae in mice is only achieved if IgM antibodies against the microfilariae are produced (26).

Some helminths masquerade behind determinants similar or identical to those of the host, to evade immunologic recognition and ultimate rejection. It has been suggested (2) that one reason why antifilarial antibody cannot be detected in microfilaraemic animals is that it is adsorbed by microfilariae. In the present study, we have shown that microfilariae of B. pahangi can adsorb immunoglobulin from the serum of their hosts, and also that with age, they are able to adsorb more immunoglobulin from the serum of the host. However, why this antibody does not harm or kill the microfilariae is not clear.

6. Acknowledgements

U.N.P. was supported by a career development grant from the Special Programme of the World Bank/UNDP/WHO. D.A.D. is a member of the external scientific staff of the MRC. The project was financed by a grant from the Tropical

Medicine Research Board and a grant from the filariasis component of the UNDP/World Bank/WHO Special Programme for Research and Training in Tropical Diseases to R.M.E.P.

7. REFERENCES

1. AXEN, R., PORATH, J., and ERNBACK, S. Chemical coupling of peptides and proteins to polysaccharides by means of cyanogen halides. Nature (Lond), 214, 1302 -1304, 1967.

2. CAPRON, A., GENTILINI, M., and VERNES, A. Le diagnostic immunologique des filarioses. Possibilites nouvelles offertes par l'immunoelectrophorese. Pathologie et Biologie, 16, 1039-1045, 1968.

3. CHAYEN, A., and PARKHOUSE, R.M.E. Preparation and properties of a cytotoxic monoclonal rat anti-mouse thyl-l antibody. Journal of Immunological Methods, 49, 17-23, 1982.

4. DENHAM, D.A. Experience with a screen for macrofilaricidal activity using transplanted adult Brugia pahangi in the peritoneal cavities of Meriones ungiculatus. Animal Models in Parasitology. The Macmillan Press Ltd., London. Edited by Dawn G. Owen. pp.93-104, 1982.

5. DENHAM, D.A., McGREEVY, P.B., SUSWILLO, R.R., and ROGERS, R. The resistance to re-infection of cats repeatedly inoculated with infective larvae of Brugia pahangi. Parasitology, 86, 11-18, 1983.

6. DENNIS, D.T. and KEAN, B.H. Isolation of microfilariae: report of a new method. Journal of Parasitology, 37, 1146-1147, 1971.

7. DEVANEY, E., and HOWELLS, R.E. The exsheathment of Brugia pahangi microfilariae under controlled conditions in vitro. Annals of Tropical Medicine and Parasitology, 73, 227-233, 1979.

8. FLETCHER, C., BIRCH, D.W., SAMAD, R. and DENHAM, D.A. Brugia pahangi infections in cats: antibody responses which correlate with the change from the microfilaraemic to the amicrofilaraemic state. Parasite Immunology, 8, 345-358, 1986.

9. GRZYEH, J.M., CAPRON, M., DISSOUS, C. and CAPRON, A. Blocking activity of rat monoclonal antibodies in experimental schistosomiasis. Journal of Immunology, 33, 998-1004, 1984.

10. de HOLLANDA, J.C., DENHAM, D.A. and SUSWILLO, R.R. The infectivity of microfilariae of Brugia pahangi of different ages to Aedes aegypti. Journal of Helminthology. 56. 155-157, 1982.

11. JOHNSON, P., MACKENZIE, C.D., SUSWILLO, R.R., and DENHAM, D.A. Serum-mediated adherence of feline granulocytes to microfilariae of Brugia pahangi in vitro: variations with parasite maturation. Parasite Immunology, 3, 69-80, 1981.

12. JUNGERY, M., CLARK, N.W.T. and PARKHOUSE, R.M.E. A major change in surface antigens during the

maturation of newborn larvae of Trichinella spiralis. Molecular and Biochemical Parasitology. 7, 101-109, 1983.

13. KAZURA, J.W., and AIKAWA, M. Host defence mechanisms against Trichinella spiralis infection in the mouse: eosinophil mediated destruction of newborn larvae in vitro. The Journal of Immunology, 124(1), 355-361, 1980.

14. KOHLER, G. and MILSTEIN, C. Derivation of specific antibody-producing tissue culture and tumour lines by cell fusion. European Journal of Immunology, 6, 511-519, 1976.

15. MACKENZIE, C.D., PRESTON, P.M., and OGILVIE, B.M. Immunological properties of the surface of parasitic nematodes. Nature, 276, 826-828, 1978.

16. MAIZELS, R.M., PARTONO, R., SRI OEMIJATI, DENHAM, D.A. and OGILVIE, B.M. Cross-reactive antigens on three stages of Brugia malayi, B. pahangi and B. timori. Parasitology, 87, 249-263, 1983.

17. MORSE, H.C., NEIDERS, M.E., LIEBERMAN, R., LAWTON, A.R., and ASOFSKY, R. Murine plasma cell secreting more than one class of immunoglobulin heavy chain. II. SAMM 368 - A plasmacytoma secreting IgG2b-K and IgA-K immunoglobulins which do not share idiotypic determinants. Journal of Immunology, 118, 1682-1689, 1977.

18. PARKHOUSE, R.M.E., and ASKONAS, B.A., Immuno-globulin M biosynthesis. Intracellular accumulation of 7S subunits. Biochemical Journal, 115, 163-169, 1969.

19. PHILIPP, M., PARKHOUSE, R.M.E., and OGILVIE, B.M. Changing proteins on the surface of a parasitic nematode. Nature, 287, 538-540, 1980.

20. PONNUDURAI, T., DENHAM, D.A., NELSON, G.S. and ROGERS, R. Studies with Brugia pahangi 4. Antibodies against adult and microfilarial stages. Journal of Helminthology, 48, 107-111, 1974.

21. SUBRAHMANYAM, D., MEHTA, K., NELSON, D., RAO, Y.V.B. and RAO, C.K. Immune reactions in human filariasis. Journal of Clinical Microbiology, 8, 228-232, 1978.

22. TANNER, M. and WEISS, N. Studies on Dipetalonema viteae (Filarioidea). II. Antibody dependent adhesion of peritoneal exudate cells to microfilariae in vitro. Acta Tropica, 35, 151-160, 1978.

23. WEISS, N., and TANNER, M. Studies on Dipetalonema viteae (Filarioidae). III. Antibody dependent cell-mediated destruction in microfilariae in vivo. Acta Tropica, 35, 151-160, 1978.

24. WONG, M.M. Studies on microfilaraemia in dogs. II. Levels of microfilaraemia in relation to immunologic responses of the host. American Journal of Tropical Medicine and Hygiene, 13, 66-77, 1964.

25. WONG, M.W., and GUEST, M.F. Filarial antibodies and eosinophilia in human subjects in an endemic area. Transactions of the Royal Society of Tropical Medicine and Hygiene, 63, 796-800, 1969.
26. WORMS, M.J., PHILIPP, M., TAYLOR, P.M., and OGILVIE, B.M. Control of _Dipetalonema viteae_ microfilaraemia in mice: Role of microfilarial surface antigens and of the antibody classes they stimulate. Parasitology, 84, xxiv, 1982.

MAMMOMONOGAMOSIS

J.EUZEBY, J.GEVREY, M.GRABER, A.MEJIA-GARCIA.

1. ABSTRACT

        In the present paper, the authors first describe two species of
Mammomonogamus of ruminants that are involved in human mammomonogamosis.
Then, they describe the features of human disease and the means of its
treatment.

2. DEFINITION:

        Mammomonogamosis is a syngamosis of mammals, due to parasitic
nematodes belonging to the Mammomonogamus genus. They affect: -ruminants
(cattle, buffaloes, sheep, goats)(an elective pharyngeal location) -Asian
elephants (pharynx, trachea) -cat (nares, pharynx) -and may occur in Man[+].
While the disease is mainly sub-clinical in ruminants, the parasite causes
a rhinopharyngeal syndrome in cats and a pharyngo-laryngeal syndrome in
man with dysphagia, chronic and vomiting-inducing cough, which cannot be
cured by usual non specific drugs.
        Mammomonogamosis in man implies a zoonotic origin, involving ru-
minants; this is the reason why the Mammomonogamus species of these animals
will only be taken into account in the present study.

3. GEOGRAPHICAL DISTRIBUTION:

        The Mammomonogamus nematodes of ruminants (and of man) are found
in tropical and sub-tropical areas of Far-East (Viet-Nam, Malaysia, India,
Philippines), America (Mexico, Central America, West-Indies, Equador,
Columbia, Argentina, Brazil, Venezuela, Gayana), Africa (Cameroon, Uganda,
Central African Republic). In these places, the infection is endemic.

4. THE PARASITES:

        Mammomonogamus spp. are nematodes, belonging to the Strongyloï-
dea sub-order and the Syngamidae family. They have a reddish colour ("red-
worms"), the males are 3-6mm long and the females 8-20mm ; both sexes are
in permanent copulation and the couple looks like a fork ("forked-worms")
(fig.1)[++]. The buccal capsule (B.C.) is cup-shaped, with a large, smooth,
mouth ring, devoided of leaf-crowns and divided into 6 festoons; the inside of the buccal capsule may be re-inforced by longitudinal ribs and the
buccal capsule bears at its base from 6 to 10 small teeth, radially-arran-
ged (fig.2). The genus Mammomonogamus is different: (1) from Syngamus and
other syngamid parasites of birds (Cyathostoma) because of the presence of
8 longitudinal ribs on the inside of the buccal capsule and of cervical
papillae. Their eggs are without opercula and contain only two cells when

+ Another genus, Rodentogamus, parasite of Rodents, does not affect man:
  neither do the avian Syngamidae.
++ Such is not the case in a genus of the same family, Cyathostoma.

laid (fig.3) (2) from <u>Rodentogamus</u> (the buccal capsule of which also has
internal ribs), because of the rounded tip of the rays of the copulatory
bursa (C.B.)(pointed in <u>Rodentogamus</u>) (fig.4). The genus <u>Mammomonogamus</u>, as
far as ruminants (and man) are concerned includes two species: (1)
<u>M.nasicola</u> (=kingi): - globular or cup-shaped B.C. - capsular ribs inequal
and of different shapes: long ones longer than half the depth of B.C. and
with the shape of french bean hull, small ones shorter than half depth of
B.C., thin and tapering; 3 ribs only are longer than half-depth of B.C.
(fig.5) - oesophagal glands short and finger-like (fig.6a)- dorsal ray of
C.B. ends in 2 or 3 branches (fig.4a) - tail of female more or less tapering and with annular superficial rings (fig.7) -(2) <u>M.laryngeus</u>: B.C. cy-
lindro-conical, with 8 internal longitudinal ribs longer than half-depth
of the capsule (fig.8) -oesophagal glands villi-like, thin and elongated
(fig.6b) -dorsal ray of C.B. undivided (fig.4b) -tail of female short and
ending in a knob (fig.9). On the other hand, the eggs of <u>M.nasicola</u> are
larger than those of <u>M.laryngeus</u>: 94 x 50 µ vs. 81 x 42 µ.
<u>M.laryngeus</u> has been found in South-East Asia, West-Indies, and Western-
coast (Pacific) of Mexico - <u>M.nasicola</u> is more frequent and has been described in South-America, West-Indies and along both western and eastern
(Atlantic) coast of Mexico.

5. MAMMOMONOGAMOSIS OF MAN

The disease is known in West-Indies (Sta Lucia, Trinidad, Porto-
Rico, and chiefly in Martinique) and mainly due to <u>M.nasicola</u>.In Mexico
both <u>M.nasicola</u> and <u>M.laryngeus</u> are involved.

The aetiology of the disease is not yet elucidated, because of the
lack of knowledge of the life-cycle of <u>Mammomonogamus</u>. We did not succeed
in infecting sheep, neither with larvated eggs containing L3, nor with
earth-worms infected by larvated eggs, nor with earth-worms picked from a
farm in which sheep harboured the parasite and the owners of which had been
affected.From the clinical point of view, Mammomonogamosis in man causes a
laryngeal syndrome: fits of dry, asthma-like, sometimes paroxystic coughing
often ending in haemoptysia and which cannot be cured by usual treatment.
Mammomonogamic cough may also lead to vomiting and, if parasites are located in bronchi, to a broncho-pneumonic syndrome.

The diagnosis of the disease is conjured up, in endemic areas, from
the above mentioned symptoms and is confirmed by endoscopy, which may show
adult worms, and by microscopic examination of sputum and faeces in which
eggs are present.

The prognosis is usually good.

The treatment may involve drawing with tweezers if the worms can be
reached, but this is usually not possible and the red <u>Mammomonogamus</u> may
not be easily visible against the red background of the mucosa. Anthelmintic treatment is easier and more reliable: benzimidazoles (thiabendazole:
50mg/kg.). - nitroxynil (5mg/kg.) and ivermectin (200 µg/kg.) might also be
tried. Since the epidemiology of the disease is not known, it is difficult
to suggest any preventive measures. Man probably become infected through
the consumption of infective larvae by phytophagy. So, in endemic areas,
it is advisable not to eat raw vegetables that might be polluted by rumi-
nant feces.

227

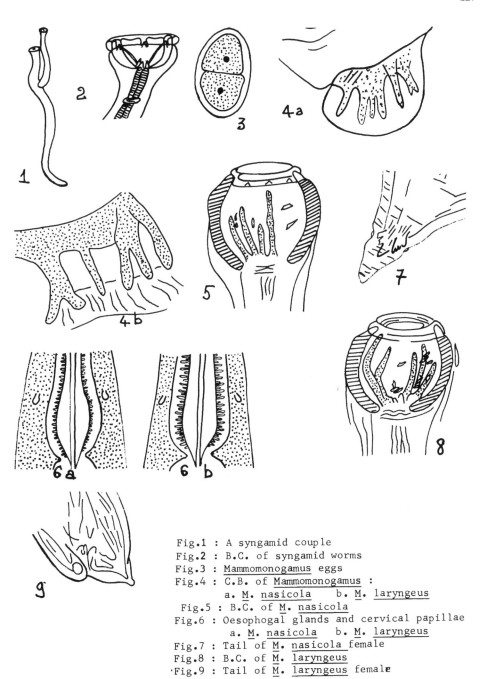

Fig.1 : A syngamid couple
Fig.2 : B.C. of syngamid worms
Fig.3 : <u>Mammomonogamus</u> eggs
Fig.4 : C.B. of <u>Mammomonogamus</u> :
    a. <u>M</u>. <u>nasicola</u>   b. <u>M</u>. <u>laryngeus</u>
 Fig.5 : B.C. of <u>M</u>. <u>nasicola</u>
Fig.6 : Oesophogal glands and cervical papillae
    a. <u>M</u>. <u>nasicola</u>   b. <u>M</u>. <u>laryngeus</u>
Fig.7 : Tail of <u>M</u>. <u>nasicola</u> female
Fig.8 : B.C. of <u>M</u>. <u>laryngeus</u>
·Fig.9 : Tail of <u>M</u>. <u>laryngeus</u> female

228

FREQUENCY OF SYMPTOMATIC HUMAN OESOPHAGOSTOMIASIS (HELMINTHOMA) IN NORTHERN TOGO

P. GIGASE, S. BAETA, V. KUMAR and J. BRANDT,

Abstract

Fifty one patients with a peculiar tumour-like abdominal condition, part of them with intestinal occlusion, were operated upon in a period of 45 months at the Centre Hospitalier Régional of Dapaong in northern Togo, near to the border with Ghana. Thirty patients were children, less than 10 years old. The M/F ratio was 1.1/1 . No geographical, social, familial or ethnic clustering was observed. The surgical appearance was highly characteristic with multiple nodules with central necrosis studding part or the whole of the large intestine. Histological appearances were typical of helminthoma in 21 of 24 biopsied cases. Twenty seven worms were observed in- or collected from the lesions of 14 patients of all ages. All identifiable specimens are immature male and female nematodes belonging to a single species, tentatively identified as Oesophagostomum bifurcum. The clinical features of the disease are recognized in the area and surgical interventions concern only part of the cases, most of which cure spontaneously in a few months. The frequency of the disease in humans and the absence of an obvious animal reservoir suggest that man himself may be the source of infection. A similar situation was described in 1964 in North-Eastern Ghana, just over the border.

1. Introduction

In 1905 Railliet and Henry (8) described as Oesophagostomum brumpti n.sp. six immature nematodes collected from tumours in the wall of the bowel of an African on the Omo river by M. Brumpt in 1902. Since then human oesophagostomiasis has been usually been considered as an uncommon disease of man, supposedly a sporadic zoonosis. In the modest number of papers dealing with this disease in humans, authors have generally refrained to express views regarding the epidemiology and the role of man as a reservoir for this parasite. We present observations to sustain the view that oesophagostomiasis might be, at least in northern Togo, a normal and rather common parasitic infection of man. The problem with oesophagostomes is that their eggs are practically indistinguishable from those of hookworms. The disease can only be identified on surgical or necropsy material and the infection on coproculture or on the identification of expelled adult worms, procedures which appear not to have been used to this purpose since more than 50 years. The failure to recognize oesophagostomum eggs has probably strengthened the impression that this worm does not normally mature in man.

2. Material and Methods

Between November 1980 and July 1984 (45 months), 51 patients with unusual large and/or painful abdominal tumours, signs of intestinal obstruction or abcesses of the bowel wall were operated by one of us (S.B.) at the Centre Hospitalier Régional (CHR) of Dapaong in northern Togo (table 1). From December 1982 to July 1984 formalin fixed biopsies of 24 cases were sent to the laboratory of pathology of the Institute of Tropical Medicine in Antwerp (P.G.). The first four specimens were routinely processed for histology and identified as human oesophagostomiasis. In the remaining specimens, the characteristic nodules were first dissected in order to collect eventual nematodes present in them. Afterwards the biopsies were proceeded for histology. Routine stains included: HES, PAS, Trichrome, Ziehl stain, Perl's stain for iron and Dominici stain for eosinophils. The collected worms were cleaned of tissue debris and preserved in 70% alcohol containing 5% glycerine. They were cleared in lacto-phenol for microscopic examination and were identified at the Laboratory of Helminthology of the I.T.M., Antwerp (J.B. and V.K.).

3. Geographical setting up

The town of Dapaong with a population of roughly 50.000 people is situated in the most northern part of the Republic of Togo (Région des Savannes), very near to the border with Ghana and Burkina Fasso. The surrounding area with a total population of 350.000 people is rather densely populated around Dapaong ($70/km^2$), but less in the southern part around Mango, where the density dropts to $30/km^2$. Several ethnic groups live in the area, mainly Moba people next to Gourma, Tchokossi and other tribes. Animists, muslims and christians are represented in about equal numbers.

The climate is Sudanese and dry. This is open savanna land, moderately arborated in the north, drier and less fertile with thorny bushes in the south. The annual rainfall is concentrated in six months from May to October and on the low side (1080 mm/year). Humidity drops to 15% during the dry season. Crops consist of cotton, the main cash crop and of millet, sorgho, beans, groundnuts, corn, rice, yams, vegetables and some fruit. Live-stock includes cattle, goats, sheep

and pigs. Guinea fowl, chickens, ducks and pigeons are reared als
poultry. Dogs are considered a delicacy in some ethnic groups and are
occasionnally eaten, only by men.
A large game reserve extends south of this area. Wild animals, i.a.
monkeys are however only occasionnally and locally seen in part of the
region where cases of human oesophagostomiasis were observed.
The CHR of Dapaong disposes of 210 beds. With the exception of the
small hospital of Mango, it is the only reference centre in the whole
area. Seven physicians were working in Dapaong and Mango in 1983,
four of them at the CHR. In 1983, 1,230 hospitalizations were recorded in the department of surgery where 498 major surgical interventions were performed. The CHR together with a moderately dense net
of health centres and dispensaries deserve the 350,000 people.

4. Results

The peculiar condition which we deal with is widely known in Dapaong
and surroundings as the "tumeur de Dapaong". Its essential clinical,
surgical and pathological findings will be briefly summarized. The
quoted name is given to more or less painful abdominal epigastric or
peri-umbilical masses which appear in a few weeks of time and most of
which disappear spontaneously or under symptomatic treatment in 6 to
12 months. Patients are generally in good condition. Fever is seldom
observed and apparently unrelated to the disease. Recent wasting was
present in 14 of the 51 patients, 10 of which suffered of intestinal
(sub)occlusion. The 51 operated patients represent only a minority of
cases occuring in this population, those with signs of (sub)occlusion
(13), of abdominal abcedation (2) or with very bulky (16) or painful
(20) abdominal tumours. At surgical intervention (fig. 1), the colon
is found to be more or less completely studded with abcedated nodules
2-3 cm in diameter, mostly on its whole length (57%) or on its right
half (31%). In one case nodules were found only on the peritoneal
serosa and omentum, not on the bowel itself. Associated localisations
were furthermore common: on the small bowel, omentum, mesentery,
liver, bladder and abdominal wall. Oedema of the wall was often
important and ascites was observed twice. The nodules may be so
numerous that no normal appearing tissue is left. The bowel loop
becomes diffusely hardened, congestive and out of shape. It is
enlarged and nodular, surrounded by adhesions. The lumen is narrowed,
but never completely obstructed and cases of occlusion have invariably
been a consequence of the extensive adhesions, usually present (80%)
around the affected bowel parts. Patients were included in this
series on the appearance at surgery of multiple subequal subserosal
nodules containing more or less inspissated material and studding
variable lenghts of the large bowel.

Table 1. Distribution of surgical cases, biopsies and cases from which worms were collected or identified with their numbers, according to age and sex

Age group	Surgery patients %			Biopsies			Worms	
	M	F	T	M	F	T	present	nr observed
1 - 9	15	15	30	6	7	13	6*	8**
10 - 19	7	5	12	3	3	6	5	11
20 - 29	1	1	2	1	-	1	1*	2**
30 - 39	2	1	3	1	-	1	-	-
40 - 49	-	2	2	-	1	1	1	1
50 - 59	2	-	2	2	-	2	1	5

* including 3 patients with worms on histology only
** including histological observations of 4 worms

Biopsy specimens from 21 of the 24 examined patients, display typical lesions. Biopsies from three patients were not representative and showed only nonspecific aspects, compatible with oesophagostomiasis. In no instance was the surgical diagnosis disproved by the histological findings. The mucosa is always intact. The lesions are spherical abscesses filled with necrotic thickened or caseous appearing material with collections of pycnotic leucocytes and sometimes crystals of Charcot-Leyden and containing either one worm or none at all (fig. 2). The wall consists of fibrous tissue with a dense inflammatory infiltrate made up of eosinophils, macrophages and plasmocytes including morular cells. In other nodes, that contain no more recognizable worms, it consists merely of dense fibrous tissue sometimes with calcifications and macrophage collections. The inflammatory reaction reaches far out from the contents and induces the extensive adhesions around the bowel. Follicular lesions were not observed. The abscesses are nearly always located between the serosa and the muscular layer, rarely in the muscle itself or in the submucosa. In this latter part lesions of another type were observed in some cases as stellate or sinuous narrow fistulous tracts filled up with fibrin, surrounded with palisadated histiocytes and a remarkably dense lymphocytic infiltrate with large germinal centres (fig. 3). They were located next to the more usual abscesses, but no transitions were observed. They are consistent in localisation and size with the passage of matured worms towards the bowel lumen, but could also represent inflammatory reactions around early destroyed larvae penetrating the wall or around collapsed superfical abscesses after death of the worm.

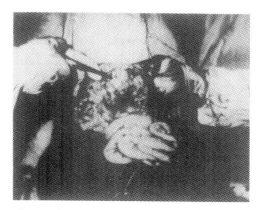

Figure 1: Appearance of the colon at surgical intervention.

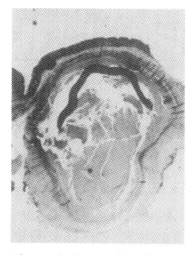

Figure 2: Composite figure to show the size of a collected female worm in relation to an histological section through the middle of a typical abscess containing a worm.

Mesenteric nodes were sometimes included in the biopsy material and found to be much enlarged with obvious germinal centre and in one specimen with granulomatous lesions consisting of epithelioid and giant cells without casation or worm rests. No stools or bowel contents were available for examination. In the lumen of an appendix which had been removed together with part of a pseudotumoral lesion, we observed sections of a mature appearing Oesophagostomum worm.

A total of 23 complete or nearly complete worms were collected from nodules in biopsy material pertaining to 11 patients (table 1). Worms were furthermore observed in histological material from 3 more patients. In any case only one worm per abscess was found. All these worms were immature adults with a M/F ratio of 1/2. A proportion of the specimens showed a sheath of third moult indicating the late fourth larval stage. The males measured 8.0-8.9 mm long and 0.38-0.41 mm in maximum width. The bursa is supported by bursal rays characteristic for the genus (Fig. 4) and the spicules measured 0.99 mm in length. The female specimens measured 10.0 to 11.4 mm long and 0.47-0.49 mm in maximum widths. The tail is tapering and measured 0.25 - 0.26 mm long. The vulva is located 0.44-0.48 mm from the tip of the tail.

5. Discussion

It is surprising that the frequency and the etiology of this condition have not been recognized earlier at Dapaong. This points probably to the relative mildness of the disease and suggests that most cases cure spontaneously without surgical intervention. It has been indeed the experience of one of us (S.B.) that such abdominal tumours respond well to symptomatic treatment with anti-inflammatory drugs and anti-biotics, notwithstanding the fact that typical nodules were still found when patients were operated after alleviation of clinical signs and symptoms. In fact, now that the etiology of the condition has been found out, surgical interventions appear unnecessary, except in those patients with signs of occlusion, abscesses or fistulisation. Histologically the diagnosis may err towards tuberculosis, especially when biopsy material consists only of caseous necrotic material and fibrous parts without specific aspects.

Figure 3: A sinuous tract in the submucosa of the large bowel.

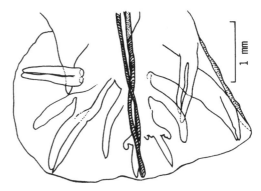

Figure 4: Drawing of a bursa of a male specimen.

The incidence of oesophagostomum infection in the area is unknown. From the age distributions of surgical cases (table 1), the infection would appear to be more common in children, but the data maybe only reflect more evident pathology in children and not a higher incidence of infection. The presence of living worms in nodules of patients of all ages favours the later view. Hitherto there are no clues as to the transmission. People of different ethnic, religious and social groups are affected. A concentration of cases was found in the rural parts immediately around the town of Dapaong, which probably only reflects easiness of access to surgical facilities. Ways of living and diet are at first sight in no way different from those of African regions with similar climatic conditions. The varied live-stock is not unusual.

The nematodes which we found are tentatively assigned to O. bifurcum, a parasite of monkeys. Morphological studies on later fourth stage larvae or on young adults are however insufficient to arrive at a precise diagnosis of the specimens.

Monkeys are uncommon in the area and their presence does not corres-pond to the clusters of cases around Dapaong. The possibility of a human reservoir can therefore not be discarded. It is possible in that case that part of the moderately frequent hookworm infestations diagnosed in the hospital represent infections by oesophagostomum. It is stressed by Togolese physicians that this peculiar and characte-ristic disease, once one is acquainted with it, is never observed in other parts of the country. The few cases seen at the Centre Hospi-talier Universitaire (CHU) of Lome, in southern Togo, on the coast, were referred from Dapaong.

In 1958 Chabaud and Larivière (3) reviewed the published cases of symptomatic human oesophagostomiasis and found records in the litera-ture of six patients, including their own, either with nodules on the bowel or abscesses of the abdominal wall. One to 187 juvenile immature worms were collected from the lesions. But for 2 cases, from Brasil and from Indonesia, all reports emanated from tropical Africa. Furthermore Leiper identified in 1911 (7) six adult oesophagostomes among hookworms passed by a Nigerian after unspecified anthelminthic treatment. In 1913 Johnson (6) claimed to have found 8 Oesophagos-tomum infections among 200 stool samples, apparently on egg morphology alone. In 1972 Anthony and McAdam recorded 22 cases in the literature but considered that the disease was much more frequent at least in Uganda (1). They presented their own series of 34 cases of helminthoma, a term which designates according to the definition of Elmes and McAdam (4) "tumourlike inflammatory swellings of the bowel, following penetration by a nematode worm". The series is based on biopsy material. Nematodes were observed nine times and could be identified in 4 cases, three times as O. apiostomum and once as Ternidens deminutus. Contrarily to our observations, the lesions were solitary in 25 cases, of which 16 occurred in the ileocecal region. Since 1972, only reports of individual cases have been published.

The paper by Haaf en Van Soest (5) deals with oesophagostomiasis of man in northern Ghana at the hospital of Bawku. In one year they observed 9 surgical cases of oesophagostomiasis, eight of which contained worms identified as O. bifurcum. More surgical cases had been observed in former years. Their findings are strikingly similar to our own observations at Dapaong, which is only 50 miles away by road from Bawku, in similar environmental conditions. In a brief discussion of their findings the authors point out that the transmission to man remains a riddle as the identified worms from man have never been described as parasites of sheeps or goats and as contacts with monkeys in this area is not sufficiently close. They suggest therefore that "the possibility that man himself may act as a source of infection cannot yet be discarded". A further case from north Ghana was published in 1979 by Barrowclough and Crome (2).

Both the Dapaong and the Bawku experiences seem to indicate a high frequency of human oesophagostomias in this subsahelian area on both sides of the border. The distribution in western Africa south of the Sahel is perhaps wider as reports from northern Nigeria might indicate. The disease is apparently well known in health centres and dispensaries in the surroundings of Dapaong and is said to bear its own native names. Many infections are probably quite asymptomatic. It is noteworthy that the worms we obtained were collected from relatively small biopsy samples cut out from the usually numerous and extensive lesions found at surgical intervention. Both the prevalence and the worm loads are therefore consistent with a human reservoir and indirect interhuman transmission in unidentified ways. The first objective of further research will be to assess the prevalence of infection in the population with appropriate means. Still, the limited extension of this high prevalence area remains unexplained as oesophagostomiasis of ruminants and monkeys is widely distributed over Africa as are sporadic cases of human infection. It may be mentioned that the only cases of pseudotumoral human oesophagostomiasis were those from north Togo in more than 22.000 biopsies from tropical Africa in the collections of the I.T.M. Antwerp, including quite a lot of abdominal pathology.

236

6. References

1. Anthony PP, McAdam IW: Helminthic pseudotumours of the bowel: thirty four cases of helminthoma. Gut, 13: 8-16, 1972.

2. Barrowclough H, Crome L: Oesophagostomiasis in man. Tropical and Geographical Medicine, 31:133-138, 1979.

3. Chabaud A, Larivière M: Sur les oesophagostomes parasites de l'homme. Bulletin de la Société de Pathologie Exotique, 51: 384-393, 1958.

4. Elmes BGT, McAdam IW: Helminthic abcess, a surgical complication of oesophagostomes and hookworms. Annals of Tropical Medicine and Parasitology, 48: 1-7, 1954.

5. Haaf E, Van Soest AH: Oesophagostomiasis in man in north Ghana. Tropical and Geographical Medicine, 16: 49-53, 1964.

6. Johnson WB: Report on entozoal infection amongst prisoners in the Zungeru Gaol, northern Nigeria. Report to the colonial office. Abstracted in Tropical Diseases Bulletin, 2: 190-191, 1913

7. Leiper RT: The occurrence of Oesophagostomum apiostum as an intestinal parasite of man in Nigeria. Journal of Tropical Medicine and Hygiene, 14: 116-118, 1911.

8. Railliet A, Henry A: Encore un nouveau sclérostomien (Oesophagostomum brumpti nov. sp.) parasite de l'homme. Comptes Rendus des Séances de la Société de Biologie, 58: 643-645, 1905

INDEX OF SUBJECTS

Artyfechinostomum malayanum, 113
Ascaris lumbricoides
 mice, immunisation, 149
Ascaris suum
 mice, experimental infection, 159
   immunisation, 149
   immunodiagnosis, larva migrans, 183
Baylisascaris procyonis
 larva migrans, review, 144
Brugia pahangi
 cat, host immunoglobulins, 215
Bulinus guernei, 122
Bulinus senegalensis, 122
Bulinus umbilicatus, 122
Cameroon
 Taenia solium, prevalence, 85
Clonorchis sinensis, 107
Cyprus
 Echinococcus granulosus, control, 60
Cysticercus bovis (see also *Taenia saginata*), 79
Cysticercus cellulosae (see also *Taenia solium*), 85, 92
Cysticercus fasciolaris (see also *Taenia taeniaeformis*), 92
Cysticercus tenuicollis, 92
Echinococcus granulosus
 camel strain, Somalia, 24
 control, Cyprus, 60
 dogs
  Tunisia, prevalence, 57
 epidemiology, review, 2
 experimental infection
  dog, 22
  monkey, camel strain, 24
  sheep, camel strain, 24
 micro- and macroecology, 2
 "Niche" theory, 7
 strain characterisation
  cloned DNA markers, 29
 strain variation, 7
 urban cycle, Tunisia, 57
Echinococcus multilocularis
 epidemiology, review, 2
Echinostoma ilocanum, 114
Echinostoma lindoense, 114
Ecuador
 Taenia solium
  chemotherapy, 103
  control, 100

Fasciolopsis buski, 111
Gastrodiscoides hominis, 112
Greece
    hydatid cyst, fertility, 12
Heterophyes heterophyes, 114
Hydatid cysts (see also *Echinococcus granulosus*)
    amino acids, kinetics, 37
    Cattle
        Greece, fertility, 12
    control, Cyprus, 60
    cyst fluid, antigen characterisation, 50
    drug targeting, 37
    goat
        Greece fertility, 12
    pig
        Greece fertility, 12
    protoscoleces
        amino acids, kinetics, 37
    sheep
        Greece, fertility, 12
        immunodiagnosis, immunoelectrophoresis, 44
Kenya
    Echinococcus granulosus, 29
Larva migrans, review, 137
Liberia
    Paragonimus uterobilateralis, 132
Mammomonogamus laryngeus
    morphology, 225
    man, disease, 226
Mammomonogamus nasicola
    morphology, 225
    man, disease, 226
Metagonimus yokogawai, 114
Nippostrongylus brasiliensis
    immunodiagnosis, larva migrans, 183
Oesophagostomum bifurcum
    man, helminthoma, 228
    morphology, 232
Opisthorchis viverrini, 108
Paragonimus uterobilateralis
    Liberia, prevalence in man, 132
Paragonimus westermani, 109, 133
Praziquantel
    Taenia solium, 104
    trematodiasis, 115
Schistosoma bovis, 119
Schistosoma curassoni
    species identification, 119
    survey, Senegal, 124
Schistosoma haematobium, 119
Schistosoma japonicum, 110
Schistosoma mattheei, 122
Schistosoma mekongi, 111
Senegal
    Schistosoma curassoni, 119

Somalia
    Echinococcus granulosus, 24
South-east and Far-east Asia
    zoonotic trematodiasis, 106
        extra-intestinal, 107
        intestinal, 111
Spain
    Echinococcus granulosus
        sheep, immunodiagnosis, 44
Taenia crassiceps, 86
Taenia saginata
    cattle, review, 68
    man, review, 68
    metacestodes
        immunoprophylaxis, review, 81
    strain variation, 76
Taenia solium
    man
        chemotherapy, Ecuador, 103
        control, 100
        metacestodes
            Cameroon, prevalence, 85
            sero-epidemiology, 85
        review, 68
    pig
        Cameroon, prevalence, 85
        control, 100
        review, 68
Taenia taeniaeformis
    immunoprophylaxis, 82
    metacestodes
        drug screening, 92
        flubendazole, 96
        mebendazole, 93
Togo
    Oesophagostomum bifurcum, 228
Toxocara canis
    larvae, excretory and secretory products
        antigen analysis, 167
        biochemistry, 168
        diagnosis, 175, 183
    larva migrans
        pathogenesis, 175
        review, 140
        role of IgG and IgE, 175
    man
        immunodiagnosis, 183
Toxocara cati, 138
Toxocara pteropodis
    larva migrans, 144
Toxocara vitulorum
    larva migrans, 143
    mice, experimental infection, 159

Trichinella spiralis
    experimental infection, horse, 192
    rodents, intestinal mast cells, 202
Tunisia
    Echinococcus granulosus, 57